W9-BOA-506

ALMOST HEAVEN

ALMOST HEAVEN

The Story of Women in Space

BETTYANN HOLTZMANN KEVLES

A Member of the Perseus Books Group
New York

Published by Basic Books,
A Member of the Perseus Books Group

Books published by Basic Books are available at special discounts for bulk purchases in the United States by corporations, institutions, and other organizations. For more information, please contact the Special Markets Department at the Perseus Books Group, 11 Cambridge Center, Cambridge MA 02142, or call (617) 252-5298, (800) 255-1514 or e-mail j.mccrary@perseusbooks.com.

Library of Congress Cataloging-in-Publication Data

Kevles, Bettyann.
Almost heaven : the story of women in space / Bettyann Holtzmann Kevles.—1st ed.
p. cm.
Includes bibliographical references.
ISBN 0-7382-0209-6
1. Women astronauts--Biography. 2. Women in astronautics. I. Title.

TL789.85.A1K48 2003
629.45'0092'2--dc21

2003013801

Text design by C. Cairl Design
Set in 11-point Adobe Garamond by the Perseus Books Group

First Edition

1 2 3 4 5 6 7 8 9 10—06 05 04 03

CONTENTS

To John and Anne-Marie Klineberg
who have shared my odyssey

PROLOGUE

IT WAS MIDNIGHT AT CAPE CANAVERAL on July 18, 1999, and still hot and humid. Even the ocean was too warm to be refreshing. The strip of sand behind the Holiday Inn at Cocoa Beach where NASA had booked its VIP visitors had been empty all day, but it would have been empty whatever the temperature. The guests, some crowded three and four to a room, had come to watch the space shuttle *Columbia* blast into orbit. Blast-offs were still public events. Scheduled for 1:30 A.M., the launch happened to coincide with the thirtieth anniversary of *Apollo 11*'s landing on the Moon. Like *Apollo 11,* this flight would break new ground. For the first time in history a woman test pilot in the United States Air Force, Lieutenant Eileen Collins, would be in command. NASA's public relations team had invited women from all over the United States to the Cape. Both NASA and the media agreed that this was a moment to reflect and celebrate women's climb to the top.

Friends of the crew, technicians and contractors who had worked on the shuttle and its payload had come from NASA bases in California, Texas, Alabama, Mississippi and Maryland to watch the launch. At the Kennedy Space Center they mixed with a huge contingent of reporters moving toward a short row of bleachers facing Banana Creek, the bay that separates the Center from its launch pads. Half a mile across the water the gigantic illuminated rocket glowed against the night sky. Earlier in the evening, the journalists had been brought to the Operations and Checkout Building to watch a prelaunch ritual—the walkout of the crew in their orange space suits as they crossed to the van that would carry them to Launch Pad 39B. Eager as they were to see the astronauts, no one pushed against the ropes separating the astronauts from the crowd. Instead they waved as soldiers protected the crew from enthusiastic well-wishers as well as potential troublemakers. Despite this show of force, the mood at Banana Creek was like a circus where the audience waited to see acrobats perform without a net.

It had been a tragic week in America. Television had not lost interest in the disappearance at sea of John Kennedy, Jr., his wife and sister-in-law in a downed plane in the ocean near Martha's Vineyard. The historical launch shared footage with the dead heir who many remembered as a three-year-old saluting his father's horse-drawn coffin.

The martyred president had started the space race almost four decades earlier. Now the Coast Guard recovered the remains of his son as the shuttle's impending launch disappeared from news broadcasts, except in Florida and Texas, the centers of America's human space flight program. Launches had become almost routine, but this one was different because Collins was the first woman to command a mission. It had taken twenty years after the first American man had rocketed into orbit for NASA to accept women, and another twenty for a woman to rise to commander. By 1999, American women had entered almost every profession, civilian and military, but this occasion was a jewel in the crown of female achievement. Collins, an aviator, would be in control of a multimillion-dollar orbiter, a remarkable vehicle that would rocket into orbit and return like a glider. She had worked hard to get to the flight deck and was ready to blast off.

Collins' five-member crew included another woman, Mission Specialist Cady Coleman, an astronomer who would help deploy the Chandra X-ray telescope, an instrument that would receive X-ray images of stellar phenomena, which are invisible to the eye; it would one day capture pictures of colliding galaxies millions of light-years away. It was not unusual anymore to have two women on a shuttle mission, but it was the first time that one of the women was in charge. Like most shuttle commanders, Collins had learned the job by piloting two shuttle missions—navigating the spacecraft and helping the commander but with no hands-on experience at the controls. Like most NASA commanders, she had earned her wings as a military test pilot. Unlike the others, she had had to break new ground as only the second person of her sex accepted into the Air Force undergraduate test pilot school in 1979, before she became the first female pilot to join the astronaut corps. Always first, she became the first female shuttle pilot when she rendezvoused with the Russian *Mir* space station in 1995.

She was married, like most of the male commanders, and like most of them she had a child. In every way she fit the profile of an American pilot astronaut, except that she was female. Unlike her male colleagues she had had to overcome sixty years of discrimination against women flying in the military and, in the years after World War II, of being banned from military jet aircraft. This shuttle mission would demonstrate that women had arrived as equal players in the Space Age.

As Collins and her crew were being checked out for the flight, Sally Ride and Kathryn Sullivan, both alumnae of the first astronaut class to accept women, talked about their own NASA "firsts." Ride was the first American woman in space, and Sullivan the first American woman to do an EVA (Extra Vehicular Activity), a space walk. Collins, about to attain another first, might become a household name like Ride, or be forgotten in just a few years like Sullivan. It would depend, of course, on Collins' performance, but it would also depend on how commonplace women astronauts became so that eventually firsts were forgotten.

If this were a one-time thing, if Collins proved to be a token, then the United States would have reneged on its promise to women that they could share in the national dream of exploring space. On this summer night in the last year of the twentieth century, a woman pilot would succeed where other women pilots had been rejected for decades. Collins had been accepted as a pilot astronaut and had made it to the top of the ladder. At the dawn of the new millennium, American women worked as firefighters, doctors and engineers and now, for the first time, a woman took command of a spaceship. A major barrier was about to be crossed, and Collins knew well that it had been a difficult journey.

NASA's public affairs department was used to providing a variety of fanfare at launches depending on the mission. The evening's carnival atmosphere celebrated the giant step for womankind. Every motel for fifty miles was booked, and latecomers bedded as far away as Orlando. The guests included First Lady Hillary Clinton and daughter, Chelsea, the entire Gold Cup–winning American women's soccer team, a selection of high-profile women journalists, including NPR's Nina Totenberg and the singer Judy Collins (no relative of the commander). Present at the personal invitation of Commander Collins were eight members of the Ninety-Nines, an elite club of women pilots founded in Valley Stream, Long Island, in 1929. (The name came from the number of women who showed up that year.)

Just after midnight, special guests, NASA technicians and engineers who had worked on the mission, and the press—all seated in bleachers facing Banana Creek, with the illuminated shuttle glowing across the water—stopped talking as the national anthem filled the air. Beyond the Center, on roads up and down the Cape, people rose to listen. A voice counted backward from ten as the crowd silently counted along. But the count stopped at six seconds to ignition. A sensor, it turned out, had indicated a hydrogen leak somewhere in the engine.

That was it. The launch was postponed and the guests informed that it takes 48 hours to restart the rockets again once the process has been aborted that late in the countdown. Word filtered through the crowd. The launch was rescheduled for two nights later. The next day the news reported that the sensor was faulty, not the engine. The launch could have taken place as scheduled.

The astronauts returned to quarters, and the launch guests found their buses and returned to their motels.

Forty-eight hours later purple and amber lightning ripped through the Florida sky presaging a storm. Rain never came, but electrical activity is just as bad. The *Columbia* launch was postponed a second time. The launch window was slim. If not that week, it would be months before another opportunity. But the next night the weather cooperated and on July 23 at 1 A.M., *Columbia* lit up the sky and the world beneath it as the towering craft lifted toward Earth orbit.

Some flights are textbook perfect. This one wasn't. Almost immediately Houston, and Collins, knew that something was awry. A liquid oxygen post pin had come out of the main injector during the main engine ignition and struck the hotwall of the nozzle, which then ruptured three liquid hydrogen coolant tubes, allowing hydrogen to escape. With less hydrogen available to combust with oxygen, the shuttle was unable, at first, to reach its targeted altitude, and fell seven miles short. Although she had piloted two missions prior to *Columbia*, Collins had never been at the controls before. She had prepared by working long hours at a simulator coping with every possible anomaly that NASA engineers could dream up. She knew, however, that she would not have a true sense of the shuttle until she was in orbit. After years of struggle, years of training, she found that handling the orbiter came naturally. Without hesitation, she oversaw several burns of the orbiter, bringing it to the appropriate altitude where the crew could launch the Chandra telescope.

They then collected data from a biological cell culture experiment and generated biological data of their own by exercising on a treadmill. Having released the largest payload in shuttle history, *Columbia* returned home, the first female commander having successfully completed her mission.

I cannot recall a time when I wasn't interested in space. As a child I fantasized about what lay beyond the Earth's atmosphere, and when space travel became a reality I thought about how wonderful it might be to be an astronaut. But unlike the women I interviewed for this book, I never considered becoming one myself.

While living in Pasadena, California, in the shadows of the California Institute of Technology and the Planetary Society, I began writing about science and technology. I met men, and a few women too, whose research and passions centered on moon rocks, X-ray telescopes and the moons of Jupiter. Several miles from my home, up in the Arroyo Seco, the dry riverbed that cuts through the Los Angeles basin, a handful of scientists and engineers had begun experimenting with rockets in the mid-1930s. Today, NASA's Jet Propulsion Laboratory (JPL) sits there several miles above the Rose Bowl.

While writing a column on science for the *Los Angeles Times,* I noticed the growing number of women rising as tenured scientists at nearby universities and as engineers at JPL. In 1983, Sally Ride rocketed into orbit, and I joined the celebration of her achievement.

As more women followed her into space, I became curious to know what had led them there. Had they been swept up by the women's movement or had they progressed along a different path? I wondered where they fit into the tapestry of ambition and possibility that so changed America as American women changed their own lives, and the lives of women elsewhere, in the last third of the twentieth century. Had women astronauts been in the lead, ahead of the curve, or had they been beneficiaries of changes that had taken place a few steps ahead of theirs? Above all, how had their participation in the astronaut corps changed NASA?

After deciding to write a book about these women, I began by asking those who had made it through NASA's selection sieve how they had gotten where they were, what kind of families shaped them, what it was like to work, or simply live, in microgravity. I spoke with the women who were part of the first group of women astronauts, all selected in 1978, some now retired from the astronaut corps, though not necessarily from NASA or from active support of the space program. Then I began visiting the Johnson Space Center, where I was able to talk to most of the women still active in the astronaut corps. They are a heterogeneous group and without exception serious, enthusiastic, intelligent and often remarkably frank.

After a conversation with Kalpana Chawla, a naturalized American from India, I realized that Americans had no monopoly on dreams of space. Explaining how her own culture both valued and devalued women, she spoke fervently of the right she believed that everyone, woman or man, has to explore as far as the imagination can soar. That led me to Chiaki Mukai, who flies with NASA as a member of the Japanese Space Agency. From there it was on to France, to find Claudie Haigneré, who happened to be in Russia at the time, and I knew that I had to go there as well. In Star City, with the help of kind Russians, now my friends, I spoke to women cosmonauts—one who had flown and others who had trained but had never made it into space. I thus expanded my search by some 6,000 miles and pushed back my story a quarter century.

From all of these conversations, in Russia, Houston and scattered places around the country, I gradually came to understand why these women had struggled to be a part of space exploration. Their voices—the women who had flown, those awaiting assignments, and those who have retired—blend to make this book a cross-cultural and multigenerational story.

In 1957, when *Sputnik* awakened the world to our ability to escape Earth's gravitational pull, no woman flew in America's Air Forces and no woman was allowed to winter over at the U.S. research station in Antarctica. Many institutions and attitudes had to change, and laws had to be both passed and implemented, before women would gain admission to our military academies, attend medical schools and become scientists and engineers whose talents NASA needed.

In 1961 the modern women's movement had scarcely begun and NASA ignored all evidence that women astronauts might be a fine idea. In the United States, the first women who struggled for a place in the male world of rocketry did not succeed. They pounded at the closed doors of space travel only to find them slammed again and again in their faces. Women in the next generation did get through, but they struggled to define their place in a male-dominated military culture. In the Soviet Union, in contrast, one woman flew in the first years of the space program, and a second before Sally Ride's historic flight.

Meanwhile, American women in the 1960s and 1970s demanded access to almost every societal niche from which they had been excluded in the past, and they insisted on getting whatever education they needed to qualify for careers that had been closed to them. Under unremitting pressure, men opened one door after another through the mid-seventies, but a few doors remained stubbornly closed. In the Soviet Union women struggled as well, but they had to work within a different set of rules.

This book, then, is the story of ambitious women whose lives were shaped by the Cold War and by the energy of the American women's movement, including the spread of its inspiration to women in many parts of the world. Very few of the women who became astronauts identify themselves as feminists; none mounted the barricades. Yet each was determined to become an astronaut, and today there is scarcely a job in the astronaut office they have not, ultimately, filled.

A woman was among the early occupants of the International Space Station, *Alpha*, which received its first crew in 2000. The ISS, always behind schedule, operates with a compromised plan and a pared-down budget. Yet incomplete and understaffed though it is, the station *Alpha* is the only place where NASA and other space agencies can learn what astronauts need to travel beyond Earth's orbit. The loss of the shuttle *Columbia,* carrying seven astronauts (including Kalpana Chawla), has changed the trajectory of human space travel and has surely delayed the next step.

A generation ago, only men walked on the Moon. When we send spacecraft to Mars, and I believe we will, there will be women in the crew. A woman might even take the first steps on its largely unknown surface. The women will be there because of the women before them. As Eileen Collins has said, "I think it is important that I point out that I didn't get here alone. There are so many women throughout this century that have gone before me and have taken to the skies."

This book is about women in space and their struggle, not altogether over, to join on an equal footing with men in what is perhaps the greatest adventure of our time. Every one of these women is remarkable, and each has bravely met two challenges—the risk of space travel, and the struggle to succeed in what was formerly a man's world. Together they have changed the cast of human exploration.

1

ASTRONAUTS AND ASTRONETTES

THE SOLDIERS PROTECTING EILEEN COLLINS and her crew in July 1999 are a reminder that, though civilian since its birth in 1958, NASA is a Cold War legacy. Created by President Dwight D. Eisenhower in direct response to the launch of the Soviet satellite *Sputnik,* NASA has always projected a military aura. This was probably inevitable, for though deliberately civilian and dedicated to peaceful exploration, many of its early leaders were military men. Their culture, not surprisingly, shaped much of the agency's attitude toward women, especially in the early years.

The launch of *Sputnik* on October 4, 1957, was perhaps the greatest technological and public relations coup of Soviet history. Lest anyone miss the satellite overhead, the Soviet news agency TASS distributed a chart, like a tidal table, that gave the precise time that the sun would strike *Sputnik*'s polished steel surface, making it visible to the naked eye. Sky watchers in London, Washington, Detroit or Bombay could know when to look up and watch, or tune in their radios to hear the signal it emitted constantly during its 90-minute orbits. Not since Pearl Harbor had America been caught so off-guard.

Unlike Pearl Harbor, of course, there were no bombs, no immediate threats to public health or safety. And there was no declaration of war. But implicit in the 90-minute orbits was a shift in power that appeared to give the Soviets a visible military advantage. Only fifteen years earlier, as America's decision-makers pondered the ending of World War II and future military threats, they had discussed, and dismissed, the possibility of building a permanent orbiting space station or a base on the Moon, or exploding a nuclear bomb on the Moon to demonstrate American strength. These proposals had an air of science fiction because the technology did not yet exist. The military chose to direct its energies toward practical, achievable aims like sending aircraft through the sound barrier or into the highest reaches of the atmosphere.

Americans had, of course, developed an atomic bomb in an extraordinarily intense, directed effort, and as a nation truly believed that given the resources, anything was possible. The Air Force had a fledgling rocket program, enriched by the help of Wernher von Braun and his colleagues—the German team who had built the rocket-propelled V-1 and V-2 bombs that had devastated British cities during the war, and who had given themselves up to American forces in 1945. But the rocket program, though progressing steadily, was not on the fast track.

Now, suddenly, space had been conquered, not by Americans but by the enemy. The Soviets let it be known that the rocket that launched *Sputnik* could just as easily carry a nuclear warhead around the world in forty minutes. *Sputnik* signaled an array of things that could come—nuclear attack, a man in space, space colonies dominated by Russians. At the very least *Sputnik* showed the world, including the neutral nations that the United States and the USSR both wooed, that the communist system had trumped the capitalists and achieved a scientific and technological triumph.

Diplomats around the world congratulated the Soviets on their accomplishment. And TASS responded by attributing the success to the legacy of Russia's Konstantine Tsiolkovsky, who, in 1903, had published an article spelling out the mathematics of orbital mechanisms and the advantage of liquid fuel for launching an artificial satellite. He had even named this satellite "Sputnik," which means "fellow traveler" in Russian. *Sputnik*'s radio batteries ran out after three weeks, but silently the bright sphere remained visible for another 1,400 orbits. Two months later, it burned up as it descended into the atmosphere.

The Soviets trumpeted their space program, which was part of their Air Force, as a military triumph. While many Americans panicked, Eisenhower did not. He knew about our Air Force program, and while acknowledging the Soviet achievement as a loss in terms of national prestige, he saw the feat as only the beginning of a duel. Eisenhower saw a public relations opportunity. In contrast to the Soviets' space program, which was military and secret, the American program would be civilian and transparent, which it was. He knew that the U.S. Air Force had an incipient space program and that we could afford two space efforts, one military and secret (which was soon largely absorbed by NASA), the other open for the world to see. In the years since he had been Supreme Commander of Allied Forces in Europe during World War II, Eisenhower had watched in dismay as the American military gained power. He would leave office in 1960 issuing a warning in his farewell address of the growth of a "military–industrial complex." With this in mind in 1958, he transformed NACA, the National Advisory Committee for Aeronautics, which had contributed to American aviation since 1915, into NASA, the National Aeronautics and Space Administration. And in the National Aeronautics and Space Act of 1958, he dedicated the agency to the peaceful exploration of space for the benefit of all mankind.

NACA left NASA a healthy budget, as well as 8,000 personnel and facilities that included three research labs—Ames in California, Langley in Virginia and Lewis (later renamed the John Glenn Center) in Ohio. The Goddard Space Flight Center was already under construction in Maryland. Almost immediately the Jet Propulsion Laboratory in Pasadena, California, run by the California Institute of Technology, joined NASA under a special relationship. Eventually NASA would establish six other centers including one at Cape Canaveral, from which it would launch spacecraft, and another in Houston for its astronauts, which would become the center for human spaceflight.

Sputnik raised the fear factor in the already uneasy fifties. These were bellicose years when the possibility of nuclear armageddon hung over the world. An Iron Curtain separated communist Eastern Europe from the West, making travel and communication between the two sides difficult and often impossible. Soviet policy aimed at worldwide communism. Under the auspices of the United Nations the democracies fought communist aggression in Korea, all the while waging a propaganda campaign to win over nonaligned countries to some form of capitalism. The Soviets promised women equal opportunities in careers like medicine and engineering; capitalists offered women consumer goods and the luxury of remaining at home to use them.

Most Americans in the fifties saw hearth and home as woman's sphere. And in reaction to the anxieties of the Depression and World War II, women married earlier than their parents and elder sisters had. They also had larger families. This was a culture that was not generous, though seldom overtly hostile, to ambitious women. Universities continued to educate women but encouraged them to use their education to raise children or become nurses and teachers, not to compete in the wider workplace. Women watching newly popular television sitcoms heard women as well as men calling them "girls" well into their dotage, subtly infantilizing all women.

Not every woman who had enjoyed interesting work during World War II retreated into domesticity. Some had slipped back into the workforce when the Korean War once again took men into the army. The draft, and the low birth rate of the thirties, produced a "manpower" shortage in the fifties, to which the Eisenhower administration responded by establishing a Presidential Commission on Women. The Commission looked at ways to utilize women in areas that traditionally depended on them—as clerical workers, teachers and nurses—and the Women's Bureau in the Department of Labor, a unit increasingly ambitious on women's behalf, explored how women could make a greater direct contribution to the nation's economy.

The Department of Labor supported protective legislation for women workers, an approach rejected by groups like the League of Women Voters, which had begun to press for an Equal Rights Amendment. This split in political objectives echoed a profound difference in how American women understood themselves vis à vis their relationship with men. At its core, this dispute revolved around whether women and men are either inherently different biologically and therefore perhaps psychologically, or basically the same with differences that are irrelevant in terms of rights. This split had existed within women's groups since women had first mobilized for their political rights in the early twentieth century, a movement that led to the adoption of the 19th Amendment, granting women the vote, in 1920. The issue of gender differences was still dividing women's organizations in the fifties, and would continue to cause rifts even as women pushed against the locked gates of one American institution after the other.

When the Korean War ended in 1953, women's demands increased incrementally. However tranquil the fifties had seemed, the voices of American women had begun to annoy some of the men accustomed to running things their own way. President John Kennedy's "New Frontier"—his campaign slogan and promise—rallied women who demanded an end to gender discrimination. In response, in 1961, he created his own Presidential Commission on the Status of Women.

Women's voices had not yet risen across the Atlantic, much less across Europe. But the Iron Curtain that divided East and West was not impermeable and ideas, even people, crossed both ways. Throughout the Cold War, international scientific organizations continued to meet under the auspices of the International Council of Scientific Unions (the ICSU), an organization established after World War I to foster international cooperation in science. The ICSU had designated the eighteen months from July 1957 to December 1958 as an International Geophysical Year, a time for nations to cooperate—and compete—peacefully.

The Soviets further stole the show with *Sputnik 2* in November 1957, a larger *Sputnik* that carried a female mongrel dog named Laika. At the many international scientific conferences that year, orbiting animals drew attention away from robotic discoveries. American scientists heard rumors that the

Soviets planned to launch a human passenger very soon, and there was even talk that they might launch a woman.

The United States launched its Explorer satellites early in 1958. They discovered several rings of charged particles surrounding the Earth that were later named the Van Allen belts after the physicist James Van Allen. But as scientifically important as that discovery proved to be, it had been preempted and overwhelmed by the *Sputnik* launches.

Sputniks in the sky and the detritus of the Korean War on the ground fueled the fires of the Cold War. A major shock to the military was the condition of the POWs who returned from Korea with the invisible scars of psychological torture that quickly won the unofficial name of "brainwashing." In response the Navy began its first studies of soldiers under stress, choosing Antarctica as a research site. The frozen continent had "hosted" scattered campsites from many nations and then, in 1959, under an international treaty that grew out of the International Geophysical Year, the entire continent was dedicated to international scientific studies.

These studies, however, were limited in most instances to male scientists. The Soviets, however, were more inclusive. In 1956 they sent Marie Klenova, a member of the Council for Antarctic Research in the USSR Academy of Sciences, a scientist who had worked near the North Pole for thirty years, as part of a Soviet oceanographic team to map uncharted areas of the Antarctic continent. There were occasional American women scientists, but they had to remain aboard ships.

The Navy controlled the research stations and until the 1969–1970 season would not permit any women ashore. Only then did the National Science Foundation, under pressure from women scientists, coerce the admiral in charge of the station—who insisted that "the Navy would be asking for trouble if women were allowed to work at McMurdo"—to allow a four-woman team to pitch their tents.

Arctic exploration moved in tandem with space exploration in holding women at arm's length. Both destinations were remote and forbidding. When Jennie Darlington accompanied her scientist husband Harry on an expedition in 1947, she shared a bunkhouse with twenty men and learned to drive a dogsled to survive. Describing her ordeal she wrote, "Imagine yourself on a spaceship in another world." That's just what scientists at NASA were beginning to do. In the interests of speed and efficiency, Eisenhower directed the newly created space agency to choose the first astronauts from a pool of men who had broken the sound barrier and flown at extremely high altitudes in spy planes, men who had already demonstrated a willingness to take risks and make rapid, life-and-death decisions: military test pilots.

From hundreds of military files, a team of NASA specialists selected a short list of test pilots whom they brought to Washington for a classified briefing. The men who volunteered for further testing were sent to a private clinic in Albuquerque, New Mexico. There they would be culled to a final group, known forever after as the Mercury Seven.

That New Mexico clinic, the Lovelace Foundation for Medical Research, was run by Randolph Lovelace, a practicing surgeon who had training as a flight surgeon, a term that applies to all military physicians. Lovelace had a monopoly on private Air Force contracts, a kind of "outsourcing" conducted not to save money but to put the findings beyond military control. Along with a medical colleague, General Donald Flickinger, who remained in the Air Force, Lovelace pursued a private medical agenda that included testing the extremes of an aviator's abilities to function under stress. As a private citizen, Lovelace had accepted the chair of NASA's new Special Advisory Committee on Life Sciences. NASA needed someone with Lovelace's veneer of independence, since the new agency avoided using the military's own clinics. This way, military volunteers who feared that discovery of a medical problem could adversely affect their careers would have their privacy protected.

The candidates NASA asked for had to be between twenty and forty years old and weigh no more than 180 pounds and have a degree in science, engineering, or the equivalent. The committee limited its search to military test pilots because Eisenhower had insisted on that. Test pilots were known risk takers, but theoretically, the committee could have considered anyone who met the criteria. Women were not specifically excluded.

NASA did not even glance at Jackie Cochran, America's most famous woman aviator, who in 1953 became the first woman to fly faster than the speed of sound, because she was too old. They did not look at younger women aviators, either, even though some had flown civilian jets. The Committee had not mentioned gender in its specs, because it was unnecessary. Although there were thousands of women in the armed services, many of whom had served in Korea, there were no female *military* pilots in 1959.

This is not to say that in 1959 there were no experienced, courageous female pilots with military experience. Women had flown *with* the American military during World War II, beginning in 1941, when women aviators made themselves available to the Air Corps Ferry Command. In June 1942 this group morphed into the Women's Auxiliary Ferrying Division (WAFS), which ferried planes and supplies to bases throughout North America. In this capacity, to protect the cargo, the WAFS were directed to remain on the ground during their menses. Although the order was never enforced—no one wanted the responsibility—there was a pervading fear that women are unstable because they menstruate, slaves of what would later be called "raging hormones," which had been associated in the past with hysteria and even insanity.

A second group of women pilots responded in 1943 when the famous pilot Jackie Cochran issued a call. Under Cochran's leadership, the Women Armed Services Pilots—WASPs—ferried cargo planes and towed targets for aerial gunners as civilian contractees. They flew risky missions and some died. But despite efforts of sympathetic congressmen to militarize the group, they never were, which meant that they did not receive life insurance or death benefits.*

With male pilots available again at the end of the war, the Army dissolved the WASP. The women who had flown with the Air Force were sent home with a simple "thank you." No women flew for the American military for another thirty years.

But women did fly. The war left a surplus of opportunities in new civilian airplane businesses, and a supply of newly trained male aviators who opened flying schools and taught women to fly. No longer a rich person's game, flying became popular among women who, while still excluded from the military, found jobs in the growing business of general aviation and in the Civil Air Patrol, as well as the continuing barnstorming kind of racing that was popular in the twenties. There were plenty of women pilots with impressive records in the fifties outside the armed services.

Their exploits were reported everywhere and must have been known to the voices in the scientific community who objected to NASA's narrow recruitment guidelines, especially the decision to limit the pool to men. Shortly after *Sputnik,* Edward Teller, known then and now as the father of the H-bomb, testified in Congress when he was asked what he thought about women as astronauts: "All astronauts should be women because they weigh less and have more sense." His testimony did not carry much force. Nor did Arthur Kantrowitz, the director of the Avco Research lab, who told the Maryland Space Research and Technology Institute in 1958 that the cost per pound of sending anyone into orbit suggested that "the first 'man' in space could well be a woman, a small woman, and probably a physician."

In 1959 Colonel John Stapp, an Air Force expert on the medical ramifications of rapid flight, reported tests showing that females functioned better than males in the cramped surroundings of a simulated 18,000-mile-per-hour flight. He also reported that females seemed to endure loneliness better than his male subjects. A Canadian team reported that their isolation studies

*Twenty-two years later in 1977 Senator Barry Goldwater sponsored legislation that recognized the contribution of these women and granted them, retroactively, military honors.

showed women enduring almost a week in an isolation chamber without experiencing hallucinations. (Men apparently weren't tested.) Later that year a psychiatrist reported in the British medical journal *Lancet* that women endured twice as long in soundproof isolation as his male subjects.

In 1959 in Santa Monica, California, ten-year-old Susan Kleinberg wrote and sent a letter to President and Mamie Eisenhower asking how she could become an astronaut. She received a reply informing her that at the moment "we are not taking little girls in the Astronaut corps." He did not add that they were not taking grown women either.

As a new agency, NASA did not initially dismiss the idea of women astronauts. For a brief time it allowed a quasi-official Air Force ARDC (Air Research and Defense Command) program that Flickinger had been working on under the rubric "girl in space" program, alternately called WISE, Women In Space Early, or WISS, Women in Space Soon. It was soon cancelled and its existence was somehow deleted from the written records. Whatever the reason for ending the program, General Flickinger seems to have transferred it to the safekeeping of his friend and colleague Randolph Lovelace at a private clinic in New Mexico.

NASA officials decided to stick with what they knew, which was how to work with men. Apparently, despite the suggestions of the aforementioned scientists, it did not occur to them that they could include women in the astronaut program without having to create an entirely separate, and expensive, women's program.

They all knew about, and some actually knew personally, the famous Jackie Cochran. But she was past fifty in 1959, and though a leader of the WASP during World War II, she was not a leader of women. Rather she was a woman who liked to set records, but as the *only* woman. With a rags-to-riches history, of which the beginnings are unknown, she capitalized on her flying fame, and married a rich man who added his wealth to hers to build a successful cosmetics company. Randolph Lovelace had first met her in 1937, when he had taken a training course as a flight surgeon and had devised an oxygen mask for her to use in high-altitude flights. In 1959 Cochran was living in Albuquerque, where Lovelace was both her doctor and friend. That same year Lovelace met Jerrie Cobb, a young female aviator as ambitious, in her own way, as Jackie Cochran was in hers.

Cobb was born in 1931 and had learned to fly when she was twelve. A devout Christian, she approached flying as a calling and by September 1959 had flown 10,000 flight-hours and held three aviation records, some in jet aircraft. As competitive as Cochran, she had received the Federation Aeronautique Internationale's Gold Wings of Achievement in civilian aircraft, an award that had won her a job as a test pilot and the rank of sales manager at the AeroDesign and Engineering Company in Oklahoma City (their first female hire).

Cobb was representing AeroDesign at an aviation meeting in Miami, Florida, in September 1959 when her boss introduced her to General Flickinger who was walking along the beach with Dr. Lovelace. Flickinger, she learned, had done aeromedical research at the old NACA and was interested in the problems of manned space flight. He had concluded that human beings could survive in space if they were extraordinarily fit physically and psychologically.

This opinion was not universal. Jerome Weisner at MIT, who would become John Kennedy's science advisor, argued forcefully to the next president that no one could survive even a few hours in orbit.

Flickinger and Lovelace had just flown to Miami that September morning from a meeting of space scientists in Moscow. They had heard rumors there about Soviet plans to orbit a man and then a woman. When they met Cobb and learned that she had thousands of hours of flight time as well as a collection of aeronautical prizes, they realized that she seemed to be just what the doctors had ordered. They arranged to meet the next day at the upscale Fountainbleau Hotel where they asked her if she would be willing to be a test subject for the first research on women as astronauts. The doctors knew that recent studies of female physiology seemed to show that women in general tolerated pain, heat, cold, loneliness and monotony as well as or better than men—which made them good candidates for spaceflight.

It seems that Flickinger's first idea was to use Cobb in the Air Force WISE program and so she first went to his Air Force laboratory for preliminary tests. When that project was discontinued, she went to the Lovelace Foundation in Albuquerque, where Flickinger asked Lovelace to give her the same tests he had given the Mercury Seven. There is some haziness about her time at the WISE program because the only evidence of the tests are letters from Flickinger explaining to Cobb that the Air Force had cancelled the program and that the tests would be continued by Lovelace at his clinic in New Mexico, if she agreed.

The Air Force, under pressure, was relinquishing most of its space programs and NASA was picking them up. Although Lovelace never explicitly told Cobb that his tests were sponsored by NASA, he hinted at the fact that the Lovelace Foundation had run the physical tests in selecting the Mercury Seven. He suggested that if her results were what he expected, he might be able to convince NASA to consider including women in the still new astronaut corps. He could not, and did not, promise her anything. Cobb, perhaps naively, interpreted his remarks as a step toward not only becoming an astronaut, an ambition she had never had before, but becoming the first woman in space.

The Lovelace Foundation, founded by Lovelace's uncle in 1922 as the Lovelace Clinic, specialized in investigating the potential and limits of the human body. It had military contracts that included monitoring the physiological reactions of pilots to the high altitudes reached in the U-2 pilots' spy flights over the Soviet Union. Lovelace's ability to remain silent about his research had earned him a reputation for keeping secrets. He had been interested in spaceflight since at least 1951, when his foundation collaborated with the Air Force School of Aviation Medicine on a conference. He had hoped to convince the government to support research in spaceflight, but the government avoided the topic, and Secretary of State Charles Wilson in the Eisenhower administration referred to the prospect disparagingly as "Buck Rogers nonsense."

Now Lovelace had a secret of his own. Cobb had agreed to the experiments, and in February 1960 she flew to Albuquerque at Lovelace's expense. He was impressed from the start with the soft-spoken woman who gave the impression that she believed she could do just about anything. Which she did for five days in his clinic, submitting to the same regimen as the Mercury astronauts. This included tests of her heart and lung capacity, her pain level and noise tolerance. She was spun in a centrifuge and tilted in an exaggerated version of an amusement park ride to test resistance to vertigo. She also had to swallow a yard of rubber tubing. She was then flown to the military laboratory at Los Alamos to have the mineral contents of her bones measured, and from there to the Veteran's Administration Hospital in Oklahoma City, where she was immersed in a pool of body-temperature water in total darkness for ten hours in a sensory deprivation experiment. Lovelace kept Cobb's high scores private.

Whatever some people at NASA knew about the tests was unofficial. But Flickinger, who sat with Lovelace on NASA's Special Committee on Bioastronautics, knew everything. When Lovelace decided to expand the study, it was Flickinger who met with Cobb to scan the list of members of the Women's Aviation Club, the Ninety-Nines, for suitable candidates. At this point Lovelace discussed the project with Jackie Cochran, and Cochran offered to help pay for the women to come to Albuquerque for a week of testing. Meanwhile, in April 1959, NASA introduced the Mercury Seven astronauts to the world.

Inspired by their success helping to select the men, Flickinger and Lovelace pressed on and selected twenty-five additional women. Lovelace asked them to keep quiet about the project, not because it was unofficial but because he did not have the means to cope with the press. He assumed, with reason, that the press would be interested—girl anything attracted attention because of the novelty—and so did anything to do with astronauts. The combination was a reporter's dream. By this time the Mercury Seven had become celebrities, long before any one of them had flown, and had lucrative contracts with *Life* magazine for their stories. But Lovelace apparently just wanted the women to keep silent for a few months because he himself revealed everything about his study of Cobb in August 1960, when he presented a paper about her performance at a Space and Naval Medicine Congress in Stockholm. He told the audience that females required less oxygen than the average male, and that women's reproductive organs, being internal, were less vulnerable to radiation. This last resounded promisingly to the assembled physicians as radiation in all manifestations was still a major issue in 1960, when many people were sensitive to the issue of radioactive fallout from nuclear testing. He also emphasized that the tests were pure research and paid for by his foundation, not NASA.

This is not the impression he had given the women pilots in his study. Indeed, the letter he wrote inviting them to Albuquerque suggested that they would be candidates in an established program for female astronauts:

> We have been informed that you may be interested in volunteering for the initial examinations for female astronaut candidates. . . . They do not commit you to any further part in the Woman in Space program unless you so desire. . . . Enclosed is a card which outlines qualifications of women astronauts.

In 1961 Lovelace selected twelve finalists from the twenty-five. By this time NASA knew all about the tests and did not want anything to do with them. It is hard to know exactly what was said to the finalists. By this time they understood that there was no women's astronaut program. But they also knew that the NASA administrator, James Webb, had invited Cobb to a banquet for the first meeting of the National Conference for the Peaceful Uses of Space and that he had asked her to become a special consultant to NASA. That sounded promising. Some of the twelve really believed Lovelace would help them become astronauts. Some did not. But whether or not they believed in the program, they were eager to go through the next phase of testing because they were all aviators and this was an otherwise impossible opportunity for them to fly the latest Navy jets in Pensacola.

Revved up and excited, the Lovelace women made life-changing decisions to get to Florida. One postponed marriage, another quit her job. None of them had grown up nursing a dream to fly in space—no one born in the thirties had. They had only just begun to dream of space travel. But for some of them the dream was overwhelming.

Events beyond their control soon marginalized their cause. On April 12, 1961, the Soviets launched Yuri Gagarin, the first human being in space. Five days later 1,400 Cuban exiles landed at the Bay of Pigs, in Cuba, expecting air support from the United States. The support did not materialize, and Castro's forces routed the exiles. With the success of Gagarin and the failure in Cuba, Kennedy went to war on the space-propaganda front. Six weeks later, in late May, he declared that the United States would land a man on the Moon before the end of the decade and bring him home safely. The space race had begun in earnest.

In August the Soviets sent up a second man, German Titov. That flight was followed five days later by a land-based affront—in mid-August construction began on a wall to separate East from West Berlin. Then in September, after an unofficial moratorium, the Soviets resumed nuclear tests. Their success in space was reverberating on the ground.

The entire nation felt enormous pressure to overtake the Soviets in space, and Lovelace's women were becoming irrelevant, a luxury the nation could not afford. Sending the first woman into orbit would of course generate national pride, but not on the same scale as landing the first man on the Moon. Two days before the Lovelace women were due in Pensacola, they learned that the tests were off. The Navy, which owned the base, had asked NASA to pay for the tests and NASA had declined.

Lovelace knew that the game was over. He accepted defeat and moved on, but not without expressing second thoughts about using women as astronauts. In 1964, in the only paper published about the Lovelace experiments, two of his staff physicians backtracked. While acknowledging the overall excellence of the women's performances, they concluded with the familiar medical canard—Yes, these women had performed well *at the particular time that each was tested.* But, they went on, women's physiology is notoriously unreliable. Fluctuations throughout the month made it difficult to test women and made it difficult to predict a woman's performance. "Monthly physiologic changes complicate . . . the women space explorer more than the male counterpart. The intricacies of matching a temperamental psychophysiologic human and the complicated machine are many and, obviously, both need to be ready at the same time."

In support of their claim, they also noted that there were studies, unspecified, showing that women who are menstruating, or near menstruating, are inattentive and accident-prone. Moreover, at such times they might experience a change in peripheral vision, especially if the woman were taking the new (and threatening) birth control pill. In a reference to word-of-mouth information from someone at the Civil Aeronautics Administration, they added that it was common knowledge that in the days surrounding menstruation women are often mentally ill and prone to suicide or crime. This "information," they reported, kept women from flying during the three days before and after as well as during their menses. For these reasons, the physicians—both former flight surgeons—did not think women should be considered as astronauts at present. Besides, since there were no women in the astronaut program, research to validate these assertions was not needed at the moment. On this last matter, they were absolutely correct.

The thirteen women who almost made it to Pensacola did not yet know each other. Each had arrived at the Lovelace Clinic alone or, like the Dietrich twins, Marion and Jan, knew perhaps one other candidate. They had not formed friendships that could be the basis for a credible opposition. With Pensacola cancelled, the candidates returned to their former life trajectories. Jerrie Cobb went back to AeroDesign in Oklahoma City. But she was not content and from Oklahoma she began lobbying for a chance for women to enter astronaut training. Assuming leadership of the group, Cobb wrote to each of the other finalists. She addressed them as "Fellow Lady Astronaut Trainees," or FLATS. The name stuck. She suggested they unite with her in pursuit of acceptance of women in the astronaut corps. This was the summer of 1961.

She thought she could win because she had a deep religious and patriotic faith. As an aviator she believed she had a God-given talent that would make her an excellent astronaut. She also believed in her nation's sense of fairness and believed that if she could reach people in high places, perhaps in Congress, she and the other women would be heard. Women had won the vote. Women now had access to education. There was a Presidential Commission on Women. Space was a new challenge, and Americans responded to challenges. She did not see why American women should be left out.

She wrote to the men she believed would help her, beginning with Robert Gilruth, director of NASA's Space Task Group overseeing the Mercury project. Gilruth delayed responding for several months and finally answered her: "We must concentrate on problems of a more pressing nature without introducing additional variables into our equation from either a scientific or public-relations standpoint." The variable he was talking about was, of course, female physiology. As for the public relations, he might have meant pressure from women's groups, or more likely the backlash he feared would arise from the accidental death of a woman astronaut.

The last of the women who Lovelace tested in Albuquerque was Jane Hart, the wife of a powerful senator from Michigan. Hart lived in Washington, D.C., and accompanied Cobb to meet Vice President Lyndon Johnson, whom Kennedy had made chairman of the Space Council. At thirty-eight, Hart was older than the other candidates, and had eight children, four boys and four girls, her own fortune, her own airplane and her own collection of aviation records. With Cobb she set out to make a case for women astronauts.

They met with Johnson on March 15, 1962, and explained to him that they were motivated by patriotism, that they believed that if the United States sent a woman into space they would be able, at last, to trump the Soviets. They believed they were as qualified as the men in the Mercury Seven program, except for the matter of military test pilot experience, from which, of course, as women they had been automatically excluded. But, they explained, each of the thirteen women they represented had logged many hours piloting private jets, some of which they had test-flown, and had done at least as well as the men in the Albuquerque tests. In short, they concluded, they felt the heavy hand of discrimination.

In fact Johnson was already worried about the homogeneity of the astronaut corps, but on the grounds of race, not gender. He wanted to see a black face in an astronaut suit. African-Americans were marching in the South, his South, demanding integration of interstate commerce facilities and lunch counters. As a Texan, Johnson was especially determined to help this constituency. Women's rights were not on his agenda. He listened politely to Cobb and Hart and made vague sounds of sympathy. His assistant Liz Carpenter suggested he could get good press out of supporting the women and wrote a letter for him to sign. Johnson not only refused but scrawled in large letters "Let's stop this now!"

Cobb waited, hoping that he would act, and then wrote to Wernher von Braun in Huntsville, Alabama. She explained to the man who headed America's rocket program that she had "talked with many people in Washington including Vice President Johnson who think it's important that the United States put the first woman in space, but unless we start soon, Russia's going to beat us again. I just keep trying, studying, hoping and praying . . . in faith."

But Cobb had heard only what she wanted to hear from the vice president and received a more polite but explicit rejection from von Braun in early May 1962. He wrote that he "appreciated" how she felt about "the embargo on lady astronauts." He, too, would like to make the trip but "am prevented by a different set of reasons." Then he added that he couldn't help "but feel that eventually you will win out. I know you would make a wonderful astronaut, or 'astronette' and will someday accomplish what you want. I think women have a place in space and eventually will assume it. However, from discussions with Mercury Management, . . . I know at the present time the thinking is that it is prudent to limit our space people to pilots with jet-test experience." He added, "I do not believe that this is a permanent thing because they are already talking about reserving the third seat in the Apollo spacecraft for a "scientist" who presumably need not even be a pilot and I don't believe of necessity must be male." He concluded with the kind of remark that was considered gracious at the time. "Besides I see no reason for not including a little beauty along with technical competence." And with those words dismissed her claims of technical competence. Cobb ignored the condescending remark—it was all too familiar at a time when she was routinely described in newspapers like the *Washington Star* as a "36–26–34 figure weighing 121 pounds"—and she moved on.

While Hart and Cobb lobbied through the winter of 1962, events continued to outdistance their cause. On February 20, NASA launched John Glenn into orbit. The United States now had its own astronaut-hero as handsome and brave as Charles Lindbergh, and as winsome as the Soviet's Yuri Gagarin. The nation was back in the game. NASA was working on the assumption that landing on the Moon first would leap over all the Soviets' smaller "firsts," including launching the first woman, which was an open secret by this time. In the battle for world public opinion, the fact that a good part of that audience was female escaped NASA's notice. Women were simply beyond their radar screen.

Jane Hart tried to remind the twenty-seven members of the House Committee on Science and Aeronautics that women cared when she wrote to them suggesting that the United States might have its own space "first" by sending a woman into orbit. Then, in June, Hart and Cobb met with Representatives George Miller of Contra Costa, California, and Victor Anfuso of Brooklyn, New York, members of the Committee on Science and Astronautics, to discuss the qualifications NASA had established for becoming an astronaut. The congressmen invited the women to tell their stories at a hearing of the Special Subcommittee on the Selection of Astronauts, which was scheduled to begin on July 17, 1962.

While they waited, Scott Carpenter became the second American to orbit the Earth, another new astronaut-hero. And in April NASA announced a call for a second group of astronauts who would fly in the proposed Gemini and Apollo programs. The medical tests would be less arduous this time and they would accept civilian pilots if they had test pilot experience. It looked good for potential women astronauts.

Then, the hearings began. Jerrie Cobb testified first. She took her time explaining in detail the remarkable performances of the thirteen women volunteers at the Lovelace Clinic. Then she turned the podium over to Hart, who got right to the point: "I am not arguing that women be admitted to space merely so that they won't feel discriminated against. I am arguing that they have a very real contribution to make. Now, no woman can get up and seriously discuss a subject like this without being painfully aware that her talk is going to inspire a lot of condescending little smiles and humorous winks. But happily for the nation, there have always been men, men like members of this committee, who have helped women succeed in roles that they were previously thought incapable of handling." She went on to describe the discrepancy between the number of women engineers in the United States and in the Soviet Union, and suggested that including women in the astronaut program would encourage American women to enter the profession.

At this point Jackie Cochran made what can only be called an "entrance." She was late and fashionably dressed as she strode to the podium. In a prepared statement that ignored the Lovelace data, she said there was not yet proof that women were as physically fit as men for flying in space. In response to a question from a subcommittee member about whether it would be a worthwhile national objective to send a woman into space before the Soviets, she waffled: "I think there are many things more important." Besides, she went on, marriage is the basic objective of all women and you waste a lot of money training them, only to have them quit. She was suggesting that *other* women could not be trusted to fulfill their missions.

Cochran's testimony startled and dismayed Cobb and Hart. Her defenders claim she acted out of patriotism, while others suspect she was a jealous older woman out to destroy younger competitors who could do what she was no longer qualified to do. It is likely that she simply resented Cobb's assumption of leadership of the FLATS. That was the role she had wanted herself.

The next day, July 18th, NASA presented its case. They led with their superstar John Glenn, who said, "If we can find any woman demonstrating better ability than the men going into this program, we will welcome them aboard with open arms." And then, according to the *Washington Post,* Glenn had to wait for the laughter to subside before he could continue. Hart had been right about the smiles and winks. Glenn went on. "I think this gets back to the way our social order is organized really. It is just a fact. The men go off and fight the wars and fly the airplanes and come back and help design and build and test them. The fact that women are not in this field is a fact of our social order. It may be undesirable."

The next to speak was Scott Carpenter, America's second astronaut. Carpenter countered Cobb's and Hart's claim that although barred from military jets, they had equivalent experience flying civilian aircraft. "A person can't enter a backstroke swimming race, and by swimming twice the distance in a crawl, qualify as a back-stroker. I believe there is the same difficulty in the type of aviation experience that 35,000 hours provides a civilian pilot and the experience a military test pilot receives."

The hearings ended that afternoon. The Committee ignored Representative James Fulton from Pennsylvania, the lone dissenter, who asked that NASA add the launching of the first woman in space to our national goals.

Cobb remained a "consultant" for a year, but was never called to consult about anything. After she had left the post, Jackie Cochran was appointed a consultant. Cochran offered to help develop a female astronaut program, like the WASP program during World War II, but nothing happened. She did please the NASA administrators with her low expectations for women and they praised her "more balanced approach to participation in the space program." When a NASA administrator tried to terminate her consultancy after eight years, she reminded him of her testimony before the House subcommittee in 1962, and stayed on.

In the wake of the hearings, Cobb received bales of hate mail from men and women alike accusing her of sabotaging the space program for selfish reasons, and by inference the entire space race and Cold War. Jackie Cochran joined the assault, describing her insistence on women astronauts as tantamount to disloyalty. Cobb did not give up hope for the long run, but for the short run she moved to Brazil, where she became a flying missionary to isolated Indians along the Amazon River.

That October, President Kennedy learned that the Soviets had missiles in silos in Cuba pointing at the United States. He would not accept this new Soviet incursion and met the Soviet premier Nikita Khrushchev head-on. The world stood at the brink of nuclear war for thirteen days until the Soviets agreed to remove the missiles. A month later, on November 21, 1962, Kennedy met with his advisors to discuss the space program. He made it clear: Space was just an arena. It was not *space* that interested him, it was the *race.* Reaching the Moon would be the ultimate test of the opposing political systems. What mattered was getting there first. "That's why we're doing it," he said. Anything that would slow the race was out of the question. And that meant the inclusion of any unknown into the process, and in 1962, one of the biggest unknowns was women.

A year later, in the summer of 1963, Martha Ann Wheatley, a reporter from the *Huntsville (Alabama) Times,* wrote to Wernher von Braun asking to interview him about women and space. The public information officer in Huntsville passed the request on to von Braun with a note: "Suggest we put Mrs. Wheatley in orbit. Would kill 2 birds with 1 stone. Seriously, I do not think we should make public comments on this highly touchy subject at this time."

So Cobb and Hart *had* made a difference. The subject had changed from laughable to touchy.

Jackie Cochran later wrote a memoir in which she reordered the historical record, placing herself at the center of the story, the way she wanted to be remembered. The FLATS kept in touch, and over time the thirteen women forged a legend of their own. Some of them believe they would have been astronauts but were rejected because of gender. Although they passed the same arduous physical examination as the Mercury Seven, none was an engineer or the equivalent and through no fault of their own they were not military test pilots. They had never, in fact, been in the running.

Although Americans became obsessed with the space race from the day *Sputnik* appeared in the sky, and had known for at least a year before that the Soviets planned to launch a woman, NASA never made an effort to best them with this "first." The suggestion of launching a Mercury woman triggered an institutional guffaw. The men at NASA were still laughing in 1963 when the Soviets launched their fifth spacecraft into orbit with cosmonaut Valentina Tereshkova aboard.

2

TWO VALENTINAS

VALENTINA PONOMAREVA NOW WORKS in Moscow as staff head of the cosmonautics group at the Institute for the History of Science and Technology at the Academy of Sciences. During the 1960s she lived forty miles to the northeast in the walled enclave of Star City, the center of the Soviet piloted space program. Every cosmonaut lived there, along with the engineers and scientists who designed and built their spacecraft, and Valentina was a cosmonaut.

In 1962, heady with the success of the *Sputnik* satellites and already preparing to launch Yuri Gagarin, the Soviets selected five women cosmonauts. One washed out before she went through training, one went into orbit, and the other three, whose existence had been unknown beyond the city walls for almost twenty-five years, were airbrushed from institutional memory until 1985, when cracks began to appear in the former Soviet Union. A journalist discovered her and her shadowy colleagues and announced their existence to the Russian-speaking world. With the total implosion of Soviet communism in 1991, Ponomareva began an excavation of the buried records of her own life, and the lives of the first class of women cosmonauts.

The Russian candidates—only Russians from the western part of the USSR were recruited—endured medical tests under conditions that simulated the anticipated stress of space travel. Like Jerrie Cobb, they were spun in a giant centrifuge and exposed to extreme gravitational forces and violent vibrations. In addition, though, the Russians had to demonstrate parachuting skill by jumping from thousands of feet. The Soviet spacecraft returned via parachute to the ground, and the cosmonauts ejected and landed with individual parachutes separately from their capsules. Like Cobb, they endured these tests in the belief that they stood a good chance of flying in space and wanted to do it for patriotic reasons. Unlike Cobb, they actually had a chance, and in the spring of 1963 Ponomareva came within a hair's breadth of rocketing into orbit.

When the possibility of space travel became real in 1961, the American women who called themselves the Mercury 13—to link themselves with the men who were chosen—and the Soviet women in the first Soviet women's group, all believed that they had been tapped in their respective nations as potential astronauts. Once they knew what it was all about, most of them longed to travel in space. This was a time when few people recognized the talents and contributions of women who did not conform to then limited ideas of what it meant to be feminine, although in both of these very different societies dissenting voices were beginning to be heard. Soviet women, like the FLATS, wanted to take part in this new adventure, but in almost every way their experiences were different. The American women never got near NASA's space center and went public with their discontent. The Russian women did reach the center of Soviet space activity but were largely silent and invisible throughout their careers.

As a child, except for the war years, Valentina Ponomareva lived in Moscow, a thriving metropolis that in the thirties celebrated a new communist order. In this self-consciously classless society she was the first born into a new kind of elite family: the technologists. Her father was an engineer, as his father had been before him. At her birth in 1933, which was followed by the arrival of two brothers and a sister, her mother, unlike most Soviet women at that time, quit work to care for her children. When the Germans invaded Russia in 1941, the family moved to the Ural Mountains for safety until they could return to Moscow in 1944.

Growing up, Ponomareva recalls, her life revolved around school. As for the influence of her parents, she explains that things were different then. Neither parent helped her plan her future; although she followed her father's example, she does not recall any special closeness with her parents. "It was a distinct trait of our generation. Unlike now, the school was home. All of our spiritual, emotional life was at school."

This was Stalin's Russia, a totalitarian society that deliberately separated children from parents to be educated by the state. It was also a society in the midst of relinquishing the utopian dreams that the Russian Revolution had promised. In 1917 the provisional government granted women the vote. After the Bolsheviks seized power in 1918 they wrote this guarantee into the July 10th Constitution of the Russian Soviet Federated Social Republic. Lenin had seen "women grow worn out in the petty, monstrous household work, their strength and time dissipated and wasted, their minds growing narrow and stale, their hearts beating slowly, their will weakened." His p. scription was the creation of institutions, like child care, that allowed womei. to work outside the home. By the midthirties women could still vote, but the Bolshevik protections—simplified divorce laws, easily available abortion and day care centers for children—had either eroded or had failed to materialize. Nonetheless, Soviet women had many opportunities to work outside the home. Indeed, most women, unlike Ponomareva's mother, were obliged to.

Women filled the Soviet labor force in greater numbers than in the United States from the beginning of the communist experiment, largely because of the devastating losses in World War I, the revolution and the civil war that followed it. In the thirties, Soviet women accompanied men on an expedition to the Arctic, and in 1935 young Nikita Khrushchev greeted an all-female team of arctic explorers on their heroic return to Moscow. During World War II, as in the West, women joined the army. However, unlike the American women who were never militarized even as they flew cargo flights for the Air Force, Soviet women flew in three women's regiments. Of particular note was the 46th Night Bomber Regiment, nicknamed the "Night Witches" for their special approach to terrorizing the Germans by flying without lights, cutting their engines as they approached the target and gliding quietly as they dropped their bombs.

Ponomareva was twelve years old when the war ended in 1945, old enough to realize that millions of men would never return home. Women, many of them widows, continued to fill men's jobs, from road workers to scientists. By 1961 almost a third of Soviet engineers were women as were nearly three-quarters of their surgeons and physicians.

In the decades after the war, Soviet scientific and engineering institutes welcomed women. Ponomareva graduated in 1957 from the prestigious Moscow Aviation Institute as an aeronautical engineer. Space travel was considered as fantastical in Russia as it was in America until *Sputnik* was launched that same year, and Yuri Gagarin flew in 1961. But immediately after Gagarin's triumphant return, the Kremlin was inundated with letters from women offering to go into space. Soviet women believed they belonged with their men in this greatest of all adventures, and unlike American men, many Soviets agreed.

While the Soviet Union was not the feminist utopia touted in its propaganda, Soviet attitudes toward women in 1960 were more egalitarian. Although American women had been able to vote since 1920, there had never been any political effort to provide publicly funded child care that would have enabled them to compete in the workplace. Although Soviet women did not have equal access to the most prestigious jobs, they did have access to important positions in medicine, engineering and management. Soviet women 'd glass ceiling, but it was higher than the ceiling in the ographs of members of the Russian space program in the like similar photos of NASA administrators, show lots of ere are a few female faces in the Soviet gallery.

et women had inherited powerful utopian promises that included establishing socialist colonies in space. Their visionaries had been reared on the literature of Konstantine Tsiolkovsky, the one-time high school teacher from rural Ryazan who, in 1903, published his formulas for rockets that could escape Earth's gravitational field. Happily for him, he was a Bolshevik, and after victory in 1917 Lenin praised him; after 1919 he received a personal pension from the Soviet government. In his elevated status he preached the inevitability of space travel and colored his vision with science fiction stories, explaining: "First, inevitably, the idea, the fantasy, the fairy tale. . . . Then, scientific calculation. Ultimately, fulfillment crowns the dream." (This was a philosophy similar to his contemporary in America, Robert Goddard, who also designed rockets and who wrote: "The dream of yesterday is the hope of today and the reality of tomorrow.") Tsiolkovsky's popularization of the possibility of rocketry spurred the growth of space clubs in the Soviet Union, where serious amateurs and professional engineers experimented with shooting small rockets into the air.

By 1931 the Moscow space club claimed over a thousand members, including a young engineer named Sergei Korolev. Within the next couple of years Korolev's club and all amateur rocket clubs were taken over by the Ministry of Defense. The Soviets had launched their plans to spread communism covertly throughout the world and tapped the enthusiasm and genius of its young rocket scientists for the job. Korolev, the heir to Tsiolkovsky's dream, would take Soviet rocketry beyond fiction and into orbit.

Born in the Ukrainian region of the Russian empire in 1907, Korolev had earned his degree in aeronautical engineering from The Bauman Higher Technical School by 1929. As a government employee throughout the thirties, he somehow survived the horrifying purges in which Stalin's secret police and armies of informers rounded up, and killed, fellow citizens for no ostensible reason. Korolev escaped execution but not arrest, and in 1938 was sent to the Gulag—prison camps in remote Siberia—where he was "rescued" from brutal labor in 1940 and brought to Tupolev's prison design bureau, a special prison for scientists and engineers in Moscow. After his release in 1944, Korolev continued to build weapons for the government and, at Germany's defeat, was involved with the capture and seizure of German rocket technology. From 1946 through Stalin's death in 1953, he worked for Beria, Stalin's henchman, and was treated as an important person.

After Stalin's death, Korolev won the attention and favor of the new Soviet premier, Nikita Khrushchev. Under Khrushchev's regime, Korolev joined the elite Soviet Academy of Sciences and designed Intercontinental Ballistic Missiles (ICBMs) to deliver nuclear warheads to the West. These were the weapons that John F. Kennedy accused Eisenhower of lacking—although the United States had at least equal power—and thus contributing to the so-called "missile gap" during the presidential election of 1960. Although there was no such gap, Americans believed there was, and Khrushchev encouraged that belief. As for Korolev, his job was to design missiles. When he could find time and governmental support, he designed *Sputnik* and then the spacecraft that took Gagarin into orbit.

By the midfifties Korolev had become the chief designer of rocket and space technologies in the Experimental Design Bureau No. 1 under the State Committee on Defense Technology. Korolev was not the only important Soviet engineer, and as part of the Soviet way of doing things, he was kept in a competitive relationship with his rivals. But after the success of *Sputnik,* Korolev rose to a special place in Khrushchev's inner circle. This did not bring him fame, however, quite the contrary. Although he was awarded the Lenin Prize and made a full academician at the Academy of Sciences, his status changed from honored citizen to valued commodity. As such, he became a state secret, his name unknown, and he was kept hidden for fear of coercion or perhaps even assassination. He was totally unknown to the general public inside and of course outside of the Soviet Union. When the Swedish Academy asked Khrushchev for the name of the designer of *Sputnik* in order to award the Nobel Prize, Khrushchev rebuffed them. It was the Soviet People, he declared, who had achieved that victory, no single individual. As Khrushchev's secret weapon, Korolev was forbidden to travel outside the Soviet Union or to communicate personally with rocket scientists in the West. Inside the Soviet Union Khrushchev kept him away from ceremonies and invisible to the public at most occasions, especially in moments of triumph. But at Star City, everyone knew and revered him. It was *his* dream the cosmonauts followed and it was he who decided which cosmonauts should fly—and that one of the first should be a female.

But deciding to fly a woman and actually selecting a woman to fly were different matters. In many ways Korolev played the role of courtier to an absolute monarch. Chairman Khrushchev was not the tyrant Stalin had been, but the Soviet Union remained a totalitarian society with a corps of secret police who could make the chairman's whim law. In this atmosphere Korolev lived with two goals: The first was personal survival, and the second, maintained since his childhood, was to put people into space. Somewhat akin to

Scheherazade's effort to entertain her king, Korolev maintained Khrushchev's interest in space by planning flights. He fed Khrushchev's lust for publicity with promises of new space spectaculars, new headline-winning missions. Although he was anonymous to the world outside the walls of Star City, Korolev was compensated in some measure by the immense respect he received inside and the opportunity he had to launch rockets into space. He acted as if he believed Tsiolkovsky's dictum: "The Earth is the cradle of mankind—but one does not live in the cradle forever."

Korolev shared his obsession with Wernher von Braun, the German rocket designer. To both, weapons were a necessary evil, a first step toward moving into space. Korolev built ICBMs in order to build a space program, and he knew that in America von Braun, who had been swept up as war booty by the American forces in 1945, was working with NASA. He sensed von Braun at his heels and determined to best him. Ironically, von Braun never knew with any certainty of Korolev's existence while Korolev was alive, cloaked as Korolev was in Soviet-style anonymity. He did, of course, learn about him after his death in 1966.

Both Americans and Russians dreamed of colonizing space. But before they could begin, they had to prove that human beings could survive there. No one thought there were dragons lurking between the planets; but there were plenty of speculations about how the absence of gravity would stop the circulation of blood or cause eyes to explode (or implode, depending). Some predicted that the neurons of the brain would fail to connect and a person would go mad. Although Korolev and von Braun believed humans could live happily outside of Earth's atmosphere, they could not be certain about all the dangers, such as solar radiation and micrometeorites, until they sent humans aloft.

The space race began, undeclared, when the Soviets selected their first cosmonauts in 1960 a year after NASA selected the Mercury Seven. The Soviets turned over their selection process to flight surgeons from the Institute for Aviation Medicine, who scanned the records of more than 1,000 male, Air Force combat pilots. They narrowed the field first to twenty and then to six for "advanced" training. As a group the Russian pilots were younger than the Americans, and slightly smaller. Slim, lean and agile, cosmonauts resembled jockeys or soccer players. Unlike the American candidates they did not have to be test pilots or engineers; bravery and experience taking risks would suffice because the *Vostok* spacecraft were highly automated. The Soviets reasoned that automation, and the cosmonaut's ultimate dependence on ground control, was necessary because no one had ever flown in space before and they reasoned that a man might lose consciousness and be unable to pilot the craft back. Worse still, they feared that the experience of being totally alone in space would drive a human being crazy.

There was a sigh of relief in both the Soviet Union and the United States when Yuri Gagarin returned sane, and when German Titov, who followed him into space four months later, was also psychologically fit. Khrushchev, as Korolev had hoped, became addicted to the routine of hugging the returned cosmonaut and parading a new Soviet hero through cheering crowds in the streets of Moscow to the Kremlin. Soviet propaganda proudly touted each flight and it was Korolev's genius to ensure that each was a first something. The *Vostok* spacecraft were all alike, but each mission extended the number of orbits, the amount of time the cosmonaut stayed in space, and from 1961 to 1963 the Soviets trumpeted each milestone: the first person to unbuckle and float in weightlessness, even the first bowel movement in space. Orbiting a woman was an obvious next step. Sergei Khrushchev, an engineer and the son of the Soviet premier who was his father's liaison with the space program, explained that it was simply part of Korolev's vision: "A man, then men, soon a woman, and then whole families." Korolev's interest in women cosmonauts was like Lovelace's. He wanted to prove that women could survive in space because he dreamed of space colonies. Ponomareva is still struck, today, by Korolev's belief that spaceflight would soon be routine and he foresaw a need to greatly expand the cosmonaut team in the near future, and to include women. This was why he was eager to find out how women would fare physiologically. Besides, Ponomareva suggests, the Soviets knew about the FLATS because "the Americans did not hide anything" and it was possible that their protests might prevail.

He had the support of Lieutenant General Nikolai Kamanin, Star City's head of cosmonaut training since 1960. As the first official Soviet hero—in February 1934 when he was just twenty-five years old, Kamanin had flown to the Arctic to rescue the entire crew of the ship *Chelyuskin* trapped in floating ice—his opinions carried weight. During World War II he had discussed the selection of women with Korolev. Kamanin respected Soviet women, and fortunately for history, he kept a diary. In 1961 he wrote:

1. *Women will definitely fly into space—thus it is better to begin training them for this kind of mission as soon as possible.*
2. *Under no circumstances should an American become the first woman in space—this would be an insult to the patriotic feelings of Soviet women.*
3. *The first Soviet [woman] cosmonaut will be as big an active advocate for communism as Gagarin and Titov turned out to be.*

A third advocate for women in space was Mstislav Keldysh, a brilliant mathematician and engineer who in 1959 headed a special Interdepartmental Scientific-Technical Council for Space Research and in 1961 became president of the Academy of Sciences. As a member of this elite academy—its membership was limited to 250—his opinions were heeded in the inner councils of the Kremlin.

It was well and good that Korolev, Kamanin and Keldysh wanted to orbit a woman, but in space matters, and in most other matters as well, final decisions were made by one man, Nikita Khrushchev. In December 1961, at his prompting—and it is not known who, if anyone, prompted him—the Central Committee of the Communist Party agreed to select a new class of cosmonauts that would include five women.

There were about as many candidates for this class as there had been men in 1959 for the class of 1960. But the selectors did not look to the Air Force, where there were no women. Instead they began by scanning the thousands of letters written in the wake of Gagarin's flight and then contacted parachute and flying clubs—both popular in the Soviet Union in the fifties and sixties. After a year of spaceflight, the selection team knew that they didn't need skilled pilots—the *Vostok* only needed piloting on reentry. (In the United States, the Mercury astronauts demanded and won a greater degree of autonomy piloting their craft.) Once back inside Earth's atmosphere, the cosmonaut ejected and released a parachute and cosmonaut and capsule floated to the ground seperately. *Vostok* cosmonauts, male or female, had to know how to handle a parachute.

With this criterion, the selection committee started with a list of 400 female candidates that they had reduced by the end of January 1962 to fifty-eight and to a final five by the beginning of March. The selection was secret but the candidates were used to secrecy. Secrecy was second nature in the Soviet Union. The women, however, were not surprised to be asked; they expected their leaders to send women cosmonauts. Four of the five finalists were chosen from either their own letters written in the wake of Gagarin's flight or recommendations from flying and parachute clubs. Only Valentina Ponomareva came recommended by the Division of Applied Mathematics at the Soviet Academy of Sciences, where she had worked in Mstislav Keldysh's lab since earning her degree. It is likely that Keldysh himself submitted her name.

She was the last woman to arrive in Star City in April 1962. Three of the other women had been there since mid-March. A fourth had arrived only the day before. They all knew that only one of them would fly first, but all hoped to get the chance to fly later. Ponomareva felt her chances to be number one were good because she was the only pilot as well as the only skilled engineer in the group.

She also seemed to hold an advantage as the protégé of academician Keldysh with a degree from the Moscow Aviation Institute. However, these matters could work both ways in a society where peasant credentials could carry more weight than academic ones. It is possible that her background was in fact a disadvantage. Her family connections, useful in the forties and early fifties, may have been the wrong ones in the sixties. On the plus side, she was an ethnic Russian, she was smart and well educated, and she had a gamine appeal, with short dark hair and a serious expression that was promising in terms of public relations.

But these qualities proved inadequate in the face of the competition. The other women were also in their twenties, bright, and either parachutists or pilots. Most had a degree from an institute of higher education, although not necessarily a technical degree. All still recall their arrival at Star City and their awe at meeting Yuri Gagarin. They do not recall any overt hostility from the male cosmonauts, but we know it was there from the diaries that some of the men have published. One referred to his female colleagues as "the female battalion before the advance." Ponomareva recalls that the cosmonauts and everyone in the military opposed the women's group and a woman's flight, but they understood that it was a matter of national prestige. It seems odd now, she says, but back then "Everyone believed that new records must be set." That meant allowing women into professions such as aviation and cosmonautics. But as individuals, Soviet men tended to be sentimental about women, especially their own mothers, and while they paid homage to ideal womanhood, on the home front they expected the real women in their lives to take care of their every creature comfort.

The arrival of the women posed multiple threats. There was the social issue of having to train with women in the same gymnasium. More basic was a matter of logistics: With few spacecraft on the assembly line and no new craft promised, the addition of even one women cosmonaut meant one less seat for a man. Beyond that was the spacecraft itself. Once launched, ground engineers were in control and the cosmonauts, wired with special medical equipment including continuous electrocardiograms in two conductors, and periodic pneumograms monitoring their lungs, became passive medical subjects. A woman cosmonaut who was not a pilot would reveal to the world how little skill a cosmonaut actually needed. Ponomareva suggests that it was in the very nature of communism to have passive cosmonauts, that mistrusting the individual is part of the ideology—the individual is small and insignificant.

Perhaps the women did not sense the coolness of their reception by the rest of the cosmonaut corps because of Gagarin's friendship and protection.

Gagarin, the son of a dairymaid and a farmer, was boyish with a warm smile and, by all accounts, a decent man and a public relations dream. He symbolized everything that Khrushchev wanted to project about the success of communism. By the time the women arrived in Star City, John Glenn had become the first American in space, and although the Soviets were still ahead, the United States had become a serious adversary and Khrushchev needed ammunition for his propaganda war.

President Kennedy had raised the stakes in that war on May 25, 1961, when he committed NASA to a Moon landing in less than eight years. In its effort to leapfrog over Soviet accomplishments, NASA ignored the other Soviet "firsts" and offered the world a contrast in style by doing everything in public. Whatever the Soviets did, they did in secret with each success announced with great flair, but only after the fact. The world did not know about Soviet failures although there were lots of rumors. Most were untrue, but secrecy bred fantasy and suspicion. The Soviet people knew only about successes when the names of orbited cosmonauts became instantly famous. Careful about these new heroes, the government anointed them, and then grounded them. The Soviet regime did not risk the lives of its heroes.

The Soviet obsession with secrecy extended to the launch site in Kazakhstan, which the Kremlin deliberately disguised on all its maps. The actual site, a town called TyuraTam, was 370 kilometers northeast of the small settlement of Baikonur, whose name was given in press releases for thirty years, although Western observers knew the true spot from tracking data. It was to TyuraTam that one of the women would travel the following year to be rocketed into orbit. At Star City the women took intense courses in astronomy and aeronautics and physical training, gymnastics and swimming, with the men. Ponomareva worried about her inexperience with parachutes. The Soviet system depended on parachuting the space capsule onto solid ground and all of the other candidates were jumpers; the youngest, Tatyana Kuznesova, was only twenty, yet already a member of the Soviet parachute team. But trouble in the pressure chamber and the centrifuge during training forced her out of the competition early.

That left four candidates. Kuznesova's friend from the parachute team, Irina Solovieva, who was twenty-four, passed all the physical tests with outstanding scores. But she was shy, and describes herself today as "withdrawn" in 1962—a poor choice, in her own eyes, to face the kind of public relations marathon that awaited the woman who would be selected. Between the last two women, both selected from parachute clubs, Zhanna Yerkina, who was twenty-two and taught foreign languages in high school, remembers being asked if she wanted to take a test to join a cosmonaut crew, but at the time she had no idea what it meant. When she asked what would happen if she passed the test, she was told, "You will be in this cosmonaut troop and you will dive from a ship." That seemed like an exciting opportunity. No one said anything about a spacecraft. The last candidate, Valentina Tereshkova, was twenty-four and alone among the four had no higher education. She had worked at a textile factory where she had an outstanding record as a leader in the Komsomol, the communist youth group—and a pedigree as humble as Gagarin's.

The four women trained from March through November 1962. They were subjected to pressure as high as ten G's (ten times the force of gravity), and they learned survival techniques for landing in water as well as on land.

When the training program ended, the surviving candidates—Tereshkova, Solovieva and Ponomareva—were formally graduated. They had been civilians until this point. Now they were commissioned as Air Force Cosmonauts. Their training continued, but now focused on a specific flight. The curriculum changed to include celestial navigation and inspirational talks from the chief designer himself. Korolev, who knew that he was too old and in poor health ever to fly in space himself, spoke to them of the devotion to duty that marks a cosmonaut's life. He suggested they forgo having children until they had made their contribution to the space program. They could be cosmonauts or mothers, he advised, but probably not both. He explained that he had not asked the same from the male cosmonauts—that their mission as cosmonauts replaced family plans. He knew of course that Valentina Ponomareva already had a small son, but he urged the unmarried women to postpone marriage and children until after their first flight.

Ponomareva, however, did not worry about her son; he was well cared for and she did not consider him a handicap. Korolev's suggestion that motherhood was incompatible with space travel did not faze her. So many Soviet mothers worked long hours that the life of a cosmonaut did not strike her as especially stressful. She did not anticipate an especially close relationship with her son in any case. The remaining contenders were all capable, but the three not chosen have said that from the first day in Star City, one woman—Valentina Tereshkova—stood out. Although her academic credentials did not equal those of her companions, her pedigree was impeccable. She had lost her father while she was a toddler during World War II, and as a young woman she worked at the same textile factory as her mother. There, as the leader of the Komsomol, she was an excellent speaker and, to top it off, as a member of a local flying club she was a fearless parachutist. The night after Gagarin's flight, she dreamed of being a cosmonaut and on impulse sat down and wrote a letter to Moscow offering her services.

In late December 1962, the fate of the three was in the hands of five men: Korolev, Kamanin, Keldysh, Gagarin and the Soviet premier, Nikita Khrushchev. Opinions differ as to who actually said what, but within a short period of time the choice was winnowed down to the two Valentinas—Ponomareva and Tereshkova. It has been suggested that Keldysh, who supported Ponomareva, may have pushed too hard for her, alienating his colleagues. Kamanin confided to his diary that while he found Ponomareva more talented, he favored Tereshkova because he disliked Ponomareva's independent behavior, which he interpreted as arrogant and self-centered. Besides, in the moralistic culture of the sixties Soviet Union, he disdained her drinking and smoking and the fact that she took walks with men who were not her husband. In Tereshkova's favor was the fact that she seemed a proven quantity, "Gagarin in a skirt." At first Korolev favored Ponomareva, because of her intelligence, but Gagarin weighed in against her. Not only did he resent Keldysh's heavy-handed lobbying, but he was especially sentimental about mothers—his mother, his wife who was the mother of his daughters—and was against sending any mother into a risky situation. Riding *Vostok* was certainly risky. Eliminating Ponomareva on principle, he opted for the other Valentina: Tereshkova.

But in the end, it was not a general or a scientist or a pilot who made the choice. It was Khrushchev. He liked Tereshkova's looks, and he liked her background. On May 21 the selections were announced: Tereshkova would fly, with Solovieva and Ponomareva as backups. Ponomareva remembers that the physician explained that instead of one backup, like the men, there were two "with the consideration of the complexity of the female organism."

Then on June 14, 1963, Tereshkova, Ponomareva, Solovieva and Yerkina watched with the rest of the Soviet public as *Vostok 5* carried cosmonaut Valeriy Bykovskiy into orbit. The plan was that for the first time there would be two *Vostok* craft in orbit at the same time, but until after his launch, the women were still uncertain as to who would be launched alongside him.

The next day Tereshkova was told she would fly with either Solovieva or Ponomareva as her backup. The day before the launch Solovieva was suited up with Tereshkova, and Ponomareva was assigned to ground support. There had apparently been some juggling of women because of their menstrual cycles, which, it seems, had fallen into synchrony—a frequent occurrence when women live together—so that they all may have been menstruating at about the same time. Whether they were menstruating or not, the launch could not be postponed because the flight, though solo, was paired with Bykovskiy. It could not be delayed.

On June 16, 1963, Valentina Tereshkova rocketed into a lifetime of celebrity, the women she had trained with fell into obscurity, their existence unacknowledged outside Star City. Ponomareva recalls sitting in the grandstand before the flight. Later, seeing a photograph taken that day, she realized that she had disappeared—she had been removed from the official record. Recounting the launch she says: "I had a motley dress on. I was standing by. After the flight I saw my dress in the news footage, but my face was not shown." She had lost a place in history, even as a footnote, until 1988. Then, in August 2002, she gave details in an interview with the newspaper of the Russian Defense Ministry. Moreover, she told them that until 1969 she was in "constant preparedness" to lead an all-female flight that had been approved by her old mentor Mstislav Keldysh and Sergei Korolev. That flight was eventually abandoned, she claimed, a victim of pressure from male colleagues at Star City.

But neither she nor the other three women lost their commissions as cosmonauts. In January 1966 Korolev died suddenly, and for a while the Soviet space program seemed rudderless as he had dominated every aspect of the human space program. The women continued training for that all-female mission, but they ignored Korolev's advice to forgo, if not marriage, then having children. Yerkina recalls: "After Tereshkova's flight all of us got married, because we could not before. We all married at the same time, except Ponomareva, who already had a child." As cosmonauts they studied and fashioned new careers, part of which was training newer male cosmonauts who would get to fly. They may not have flown themselves but their inclusion in the cosmonaut corps provided them with rare privileges, educational opportunities and status in the Soviet Union.

While her friends and colleagues watched that June morning, and with Bykovskiy still in orbit, Tereshkova donned an orange spacesuit with a helmet attached and climbed into her own spacecraft, *Vostok 6,* to join him in space. She was the sixth Russian, the eleventh human being, and the first woman in space.

Within minutes of her launch, the cosmonauts established radio contact with each other. Separated by five kilometers, they were closer to each other at that time than they would be for the rest of their flights. From their flight paths, it was impossible the Bykovskiy ever glimpsed Tereshkova's face through the small glass window in the spacecraft. Yet, he reported later that he could tell she was ill. Others saw her live on television from inside the craft within three hours of the launch. To some she looked well and happy, to a few in ground control she seemed wan. To her mother, who only discovered what her daughter had been up to when she saw the pictures on a public television screen in her hometown, Valentina looked wonderful.

What we know about Tereshkova's experience comes from Soviet press releases and her carefully crafted memoir and essays. She had been given the code name "Sea Gull," perhaps a bow to her femininity, as the male cosmonauts who preceded her had the names of predator birds like "Falcon," Bykovskiy's code name. Whether she chose Sea Gull or was assigned it, she liked the bird, which was known for its beauty, and she has used "Sea Gull" as her talisman ever since. Once in orbit she identified herself proudly to Bykovskiy: "It is I, Sea Gull." After this exchange, she seems to have communicated sluggishly with ground control. Moscow suspected she was suffering space-induced nausea as Titov had two years earlier. Tereshkova insisted she felt fine.

Fine but not comfortable, as she wrote frankly in her notes. She wore a very uncomfortable helmet fitted with electrodes that were supposed to monitor her mental state, and she apparently refused to take the pill she was asked to take or, for that matter, to participate in any of the medical experiments she was trained to do. This prompted ground control, fearful of her well-being and ability to pilot the capsule back, to consider bringing her in ahead of schedule. But she rejected the offer, insisting that she felt fine and wanted to carry out everything she had been prepared to do. That meant spending three days in orbit.

On the second day a crisis arose when she failed to perform a major goal of the mission—manual control of the spacecraft. Korolev worried that she would be unable to work the altitude control system should the automatic reentry system fail. Kamanin was worried, too, and he ordered Gagarin, Titov and Nikolayev, his experienced cosmonauts, to instruct her from the ground on manual orientation. Finally, during her forty-fifth orbit, she was able to gain control of the spacecraft manually for fifteen minutes.

She kept orbiting, but in addition to not performing all the medical experiments she had been assigned, she failed to keep a log, filling in the details only after landing. She did manage during her three days in space to take lots of movies, but here again, not smoothly. She had trouble getting the film out of the camera. She found weightlessness merely "not unpleasant," in contrast to Bykovskiy, her companion in space, who unbelted himself five times during his five days in orbit and floated for an entire orbit, enjoying the novelty of the experience. She missed a toothbrush, which she said added to her feeling of personal distress. In addition the helmet gave her a constant headache, and she complained that the sanitary napkins, whether used as a diaper or for her period, were too small.* Finally, she disliked the food, found the bread dry and attributed her one admitted episode of vomiting to the food rather than to any inner ear disorder.

She remained silent throughout reentry, not reporting on the retrofiring or the separation of the modules, which frightened the ground controllers trying to monitor her progress. She ejected from the capsule at the right time, but there, too, she violated procedure by looking to the side of her parachute instead of staring straight ahead as carefully instructed. This made her vulnerable to a piece of metal that hit her flat on the nose as she left the capsule. She landed safely on June 19, several hundred miles northeast of the town of Karaganda in Kazakhstan. A group of villagers crowded around, eager to greet her, and at this point she apparently gave them all of her remaining food so that it was impossible to measure what she had or had not eaten. Her flight was not flawless, but none of the first flights were. She wrote an account soon afterward, and an expanded account later explaining her actions.

Undisputed in the account of her flight was the reaction of the people who found her. They believed she was starving because she wolfed down a thick slice of black bread and salt. Whether she ate out of hunger or graciousness, she was quick to put the right spin on the episode. She explained that she was not starving but welcomed the custom of greeting a guest with bread and onion. Her explanation delighted Khrushchev and the propaganda machine in Moscow.

Not so her colleagues at Star City. An undeclared contest, still unresolved, followed between Tereshkova, who was never forthright about her experiences, and the physicians who prepared a hypercritical report accusing her of a weak mental and physical performance. One doctor suggested that she had been on the brink of a mental breakdown or, at the least, of being extremely secretive. Tereshkova said nothing for weeks while engineers searched for an explanation as to how the glass in the window of the capsule had cracked. This is a strange accusation as the *Vostok* hit the ground hard, which is why the cosmonauts bailed out in parachutes to ensure a soft landing. Her critics insisted that such a weakness in the glass could be dangerous and had to be explained. Finally, after weeks of silence, she volunteered that she had broken the window with her camera when she was having trouble removing the film cartridge.

News of the conflict reached Korolev, who invited Tereshkova into his office for a one-on-one chat. The outcome seems to have so dismayed

---

*Tereshkova writes of sanitary napkins in addition to arrangements for collecting urine. It is interesting that this possibility of her menses did not scuttle her flight. A decade later, when NASA discussed sending women, they were concerned about the danger that menstruation might pose for an orbiting female.

Korolev, according to Kamanin's diary, that the Chief Designer (as Korolev was always referred to) swore never to "get involved with broads again."

But why dismiss all women if, indeed, this one woman's performance—a woman he knew to be younger and less educated than the other choices—was flawed? Why pick on her at all? Tereshkova was not the first cosmonaut to obscure the truth. Her colleagues recall that the Soviet Union in those days was a sea of innuendoes, whispering and secret spies, with the Gulag only a betrayal away. When asked about Tereshkova's silence, Zhanna Yerkina, one of the four women trainees, explained: "At that period, it was better to remain silent. It would have been unwise to report on [her condition]. She was keeping herself very private." She was also, in a sense, following the lessons that the first group of cosmonauts had taught the women's group, "how to deceive physicians and how to pass tests easier." Ordinary Soviet citizens often manipulated the truth about their lives. Cosmonauts did too, and so have some American astronauts. We now know that temporary nausea affects about 80 percent of all people in their first days of microgravity. In the early sixties it was not yet named—it's now called Space Adaptation Syndrome—much less acknowledged.

Tereshkova defended her performance. She explains in her memoir that she carried out the physics experiments she was trained to do and placed the failure of the biomedical experiments at the door of the design of the spacecraft, which made her unable to reach the equipment. She has never explained the problems with handling the spacecraft or her silence during reentry. Whether the faults were all hers is not known. We do not know if those who selected and trained her ever questioned their decision. It did not matter to Khrushchev, who was pleased with Tereshkova and did not care about sending another woman into orbit. Even a decade after the end of communism her female colleagues, including those who might have been more successful, defend her performance. They too suffered when the decision-makers in Star City attributed all of Tereshkova's shortcomings to gender. In an interview in 1999 Dr. Olga Kozerenko, who started working with cosmonauts in 1961, recalls very clearly that Tereshkova performed excellently in all the pre-flight tests and did equally well during her flight. According to Kozerenko, Tereshkova performed so well that she answered all the questions about the ability of women to withstand the pressure of spaceflight, which is why no other woman was sent up for almost two decades. Whether she was superb or a failure, no one blamed the very idiosyncratic selection process for whatever shortcomings Tereshkova may have demonstrated. It is possible that Valentina Ponomareva, with her experience and education, would have been a better choice scientifically. But Tereshkova was a resounding success as an icon of female achievement.

Besides, the accusations and recriminations about Tereshkova did not travel beyond the walls of Star City. In Moscow she was awarded the honors of Hero of the Soviet Union and the Order of Lenin, To the world at large, Valentina Tereshkova was a hero, and Khrushchev wallowed in the goodwill she generated. Soviet women were touted as the equals of men, and Tereshkova was elevated to the rank of first among equals. Her oratory talent was considerable, a gift to Khrushchev, who sent her to speak throughout the Soviet Union and around the world.

Four months after her flight, Khrushchev hosted the first state wedding in Soviet history and personally gave away the bride, Valentina Tereshkova, to Andrian Nikolayev, the cosmonaut who had preceded her on *Vostok 3*. Several months later she produced a daughter, proof that space travel did not interfere with love, fertility and the birth of a healthy child.

Nineteen sixty-three was Tereshkova's year. Despite the disdain of some of her male colleagues, she remained close to Yuri Gagarin and spread the message that Soviet women could do whatever their male counterparts did. Tereshkova remained in Star City as a cosmonaut and earned a degree in aeronautics, always hoping to return to space, offering the press news of another assignment that never came to pass. In 1982 Tereshkova quietly divorced Nikolayev and remarried. Like Gagarin, who died in a plane crash in 1968, and Titov, who lived until 2000, Tereshkova never flew again.

Khrushchev had been right about Tereshkova as a saleswoman for communism and he used her constantly, as did his successors, in the first decades after her flight. Eventually she earned the reputation of a prima donna, accepting invitations and then changing the date at the last minute, or not appearing at all.

Neither she nor her cosmonaut colleagues won much admiration in the United States in the years before Apollo. America's astronauts disparaged Soviet space travel. Because so much of the *Vostok* system was automated, they later described the cosmonauts, all of whom were professional Air Force pilots except for Tereshkova, as "spam in a can." They parlayed the phrase into negotiating a place for pilot controls in American spacecraft, so they would not seem useless, as the Americans caricatured their Soviet adversaries.

Tereshkova's flight may have been intended as the first of several women's missions. Korolev had talked about flights with perhaps two of the women in orbit together. Even after his sudden death in 1966, the women kept training. Ponomareva recalls Gagarin, who reigned as a charismatic hero at Star

City, saying that he could not imagine her group being disbanded, for then where would Tereshkova go? "She would be alone without a group," a condition that Gagarin, for one, found unthinkable. Another reason, Ponomareva suggests, was that although the commanders of the Center did indeed want to get rid of them, the fact that they had become regular officers made it difficult. So the women kept training for flights that were always, for one reason or another, cancelled. Although in the end only Tereshkova flew, she was not intended to be a token female, the only female cosmonaut for a generation. As for the others, though they were out of the picture, they were not abandoned. They continued working with the space program and had successful careers as a psychologist, engineer and eventually a historian. Valentina Ponomareva lobbied, within the limits of a totalitarian society, for a chance to fly. She could not, as Cobb and Hart had, take her case to the public. She knew from the beginning of training that only one of them could be first. It was reasonable to hope that she might be the second, third or fourth woman in space.

In June 1963 Tereshkova's photo dominated the front page of American newspapers. Exactly a year earlier, the House Committee on Science and Aeronautics had dismissed Jerrie Cobb and Jane Hart after less than a day's hearing. Now reporters pressed the two women to comment on Tereshkova's flight. Cobb, a gentle woman, answered as if wondering aloud: Why had the United States let the Russians have this "first"? Hart, a more acerbic personality, told the press that she had been "tempted to go out to the barn and tell the whole story to my horse and listen to him laugh."

Most Americans were more anti-Soviet than pro-women, and many who acknowledged this particular milestone resented the society that had pulled it off. In 1963 Americans feared Soviet missiles and found everything about the Soviet system inherently loathsome. Around the United States it was commonplace to read that it was "typically Russian" (few Americans used the term Soviet) to make a woman endure the rigors of space for what all agreed was just a publicity stunt, a performance without any scientific or even historical value.

The most gracious recognition of Tereshkova's flight appeared in an editorial in the *New York Times:* "The Soviet Union has given us the first heroine of the Space Age, thus demonstrating that its long lead in the science of space travel is matched by its acumen in space showmanship. . . . The sad part is that there must be any nationalistic tinge to this achievement. . . . To Valentina Tereshkova, who went from spinning cotton in a textile factory to spinning through the heavens, . . . go our hopes and prayers for a successful journey."

Sara Gibson Blanding, president of then all-female Vassar College, told the *New York Times* that "the pattern has been that the Russians have made opportunities available to women more than we have." But she had confidence in America. "I'm sure we'll follow along in sending a woman into space." Of course she was right, but she may not have thought that "soon" really meant in twenty years. The women's movement had barely begun in the United States, and some women found feminist demands both disorienting and frightening. A female columnist in the *Orlando Sentinel* warned "girls clamoring for equal rights as astronettes" to consider the problems of hygiene in space. No woman, she asserted, would want to wear the same space suit without a bath or change of clothes for six weeks, not to mention the lack of deodorants and moisturizers—a lack, she said, that would not bother boys.

Turning to experts, the *New York Herald Tribune* asked the venerable anthropologist Margaret Mead to interpret Tereshkova's flight, and Mead responded: "Russians treat men and women as interchangeable whereas we treat men and women differently." If Americans were to send up women, she explained, it would be "because women may see things that the men might not see. We train women to look at the human factor and men to look at things."

Betty Friedan, who had just published *The Feminine Mystique,* the book that triggered middle-class American women's drive for equality, disagreed with Mead about the nature of women and men. She did not agree that women look at the human factor while men see objects, and she certainly objected to the idea that there was "a special feminine mystique that women would bring into space. . . . It's really kind of sad the way people in America look at the Russian achievement as something a little obscene." Friedan was caught in a bind between her own personal history as an anti-fascist socialist, which she did not want to publicize, and a desire to defend the Soviet achievement because they had, after all, launched a woman, even if maybe it was a geopolitical stunt. She was not naive enough to believe that the Soviet motive was to send a woman who would "see things that the men" might miss. Who could say why the Soviets had launched a woman? Not Friedan, but she used the occasion, as she used every occasion, to note that the United States was "so far from really seeing women in terms of their full potential that to send up an American woman would seem to be a publicity stunt."

Clare Booth Luce, a playwright and former Eisenhower ambassador to Italy, unabashedly admired Tereshkova: "Once again the Russians have shown that they knew how to get ahead of us by letting women assume an equal share in society. We must stop trying to make of women paper dolls. They don't like it. They don't need it." In a full-page essay in her husband's publication, *Life* magazine, she called the failure of American men to send the first woman into space perhaps "their costliest Cold War blunder." She continued: "The Communist system confronts the American system in a life-and-death struggle. Russia's men know it; so do her women. . . . It is against this background of the participation of Russian women in every effort, from sweeping the stables to combing the stars, that we must view the flight of the first woman cosmonaut. . . . The astronaut of today is the world's most prestigious popular idol . . . [and] Russian women . . . actively share (not passively bask, like American women) in the glory of conquering space."

For some men at NASA women were not even paper dolls—they were an embarrassment. An unidentified NASA official spoke for this constituency, declaring that the talk of an American space woman "makes me sick at my stomach."

NASA had known for at least three years that a woman cosmonaut was on the Soviet agenda, but the Moon missions, known as Apollo, completely filled NASA's field of vision. NASA never considered preempting the Russians by launching an American woman astronaut because at the time they could not find a woman acceptable to the Mercury program. Besides, while Korolev dreamed of colonies with men, women and children in space, NASA's dream was more pointedly aimed at reaching the Moon. Tereshkova's flight elicited sneers from men like General Leighton Davis, commander of the Air Force Missile Range at Cape Canaveral, who joined the chorus of critics in dismissing her record-setting flight as "merely a publicity stunt." He might have noted that the great effort to land an American man on the Moon wasn't much different.

3

STILL GROUNDED

ALONE, TWISTING SLOWLY IN MICROGRAVITY, a nubile and scantily clad astronaut snaps to attention as she hears herself paged.

"Barbarella!" The President of the Republic of Earth in the Sun system gets her attention. His bland face fills the video monitor as Barbarella, half nude, creeps across the fur-lined interior of her spacecraft to face him and learn her assignment. Her task is simple enough, she finds: eradicating the evildoers on planet Tau. It is the forty-first century and the astronaut Barbarella is the first American woman to command a spaceship.

The film *Barbarella,* adapted from a serial in a French magazine, was released in 1968, five years after the Soviets orbited cosmonaut Valentina Tereshkova. Barbarella resembles the gutsy Russian, an almost-feminist sex-kitten superwoman. Like Tereshkova, Gagarin and each of the Mercury astronauts, Barbarella is alone in her spacecraft. Like them, she is doing something brave for her country and for all humankind.

Nineteen sixty-eight, the year *Barbarella* appeared, women demonstrated against the Miss America pageant in Atlantic City by tossing bras and girdles, "instruments of torture," into a trash can. Barbarella is part of this scene, embodying the schizophrenic image of its young women. She is not yet a self-conscious feminist, but neither is she a traditional American girl. Barbarella discovers that the villain on Tau is a woman. In fact, all the strong characters in the film, good and evil, are female. No pacifist, Barbarella destroys her enemy to save the universe. Although women generally ignored *Barbarella* in 1968 when it was marketed, with reason, as soft porn, it is a feminist film. Barbarella is sexually liberated and an astronaut, a participant in the adventure of space travel from which American women, in 1968, were excluded.

Unlike in *Barbarella,* Stanley Kubrick did not project much female power in his *2001: A Space Odyssey,* also released in 1968. He nods to a few women scientists but still has a stewardess serving meals on his spacecraft. Young women in 1968 read science fiction, especially Madeleine L'Engle's *A Wrinkle in Time,* whose protagonist is female, but they were as far from finding a place on a spaceship as Jerrie Cobb. And along with their brothers, they kept one eye on the Apollo flights, and the other on the war in Vietnam. Lyndon Johnson, who as vice president under Kennedy had overseen the space program, succumbed to pressure from anti-war activists and chose not to run for reelection. Fighting intensified in Vietnam, and for the first time Americans watched from easy chairs in their living rooms as their soldiers were lifted out of the jungle in body bags. Still glued to television a year later, in July 1969 they watched astronauts Neil Armstrong and Buzz Aldrin walk on the Moon. Even in black and white, it was truly awesome.

In 1968, as movie cameras depicted a fantasy feminist, a group of real, down-to-Earth women were organizing a lobby in Washington that would change the face of American politics. It would also ensure that within the next decade, the United States would have real female astronauts. Feminism had never disappeared from the American political scene, but in the post-World War II years it had fragmented into small groups operating largely within the labor movement. In the early sixties, feminism seemed to explode out of nowhere, but of course what it really did was focus the discontent of a generation that had not known feminism as a political movement. The manifesto of the movement was Betty Friedan's *The Feminine Mystique,* which gave voice to the frustration of women locked into a domesticity that offered little satisfaction. Friedan did not appeal to all women, but she struck a chord with enough women to rally them to the next step—political change.

In 1964 American women had been earning 60 cents for every dollar earned by men, and they accounted for less than 1 percent of federal judges, less than 4 percent of lawyers and just 7 percent of doctors. In 1964, energized by a newly articulated sense of injustice, women in Congress watched Virginia senator Howard K. Smith add the word "sex"—as a joke (some scholars say that he was serious)—to Title VII of the Civil Rights Act, which was a major administration priority. When the bill was about to pass, a committee that tried to remove the offending word was stopped in its tracks when a congresswoman and a female senator threatened to scuttle it all unless the word remained. They won. The word "sex" was added to the clause in Title VII that prohibited businesses with more than fifteen employees from discriminating on account of religion, race or ethnicity. But the battle was not yet won because the new Equal Employment Opportunities Commission

(EEOC), created to enforce the Act, refused to enforce that phrase, which its director called "a fluke conceived out of wedlock."

It was a tactical error because Betty Friedan, with her book a success, had become a political activist. In 1966 she was in Washington for discussions about the Equal Rights Amendment (ERA) when she joined a group of about thirty people in the community meeting room at the *Washington Post.* They had gathered to charter the National Organization for Women (NOW), which lobbied for women and helped transform the national political scene. Among NOW's founders that day in Washington was Jane Hart, the alumna of Lovelace's experiments who, along with Jerrie Cobb, had suffered dismissal by the special subcommittee on astronautics in 1962 when they presented their case for women astronauts. In less than ten years NASA bowed to what seemed inevitable and recruited women astronauts.

The FLATS had cheered NASA's effort to get its men to the Moon, but some of them surely grimaced when they read in *Parade* magazine, a Sunday newspaper supplement, the response Wernher von Braun had given after a speech at Mississippi State College on November 19, 1962, to someone asking if NASA planned to use women astronauts. He had replied: "Well, all I can say is that the male astronauts are all for it. And as my friend Bob Gilruth says, we're reserving 110 pounds of payload for recreational equipment." *Parade* ran von Braun's picture along with the text. Its publication triggered so much criticism that the public relations office at the Marshall Space Center, where von Braun worked, wrote, but never mailed, a response in von Braun's name listing the number of nonastronaut women who worked in the space program. The absence of women astronauts was beginning to embarrass some people at NASA, especially those involved with selecting astronauts.

In 1965, at least one NASA official thought highly enough of women pilots to ask Janette Piccard, a famous Belgian balloonist and aviation hero, to consult at the Marshall Space Center. He wanted her to talk up NASA's accomplishments. "You're famous, the people still remember what you did, and it can help our program," he told her, revealing that a NASA administrator knew that the space program needed popular support and felt that a famous woman aviator could help. Perhaps he sensed that her support would win women voters to the cause of space exploration.

All well and good to have female boosters, but nothing was said about helping women who wanted to join the party. The lack of women at NASA, not to mention women astronauts, inspired the president of all-female Mills College in Oakland, California, to celebrate its 114th Founders Day on October 28, 1966, with a symposium: "Women's Place in Space Science."

The subject was *not* women's place in space, but it was close. The speakers included two women scientists and the female personnel officer from NASA's Ames Research Laboratory in nearby Mountain View. There were three other women speakers: Marjorie Slayton, Josephine Schirra and Louise Shepard, wives of three of the Mercury Seven. The panel was moderated by Caspar Weinberger, then head of the California Republican Party, and other speakers included astronaut John Glenn, NASA's administrator James Webb and Congressman George Miller, a Democrat from Alameda, California, who sat on the House Committee on Space and Astronautics.

Soviet space competition was topmost in the mind of the Ames' personnel officer. She took the opportunity to point out that only 1 percent of American engineers were women, whereas in Russia the percentage was 20 percent. She explained that in the United States there were strong cultural pressures on women not to compete with men, but she reminded the female student body that women could choose marriage and a career. She added a caveat, referring to her own experience, that they would do well to choose "understanding husbands."

Congressman Miller was the only speaker who thought that women should actually go into space. He acknowledged that women could not become astronauts just then because there were no women test pilots. So he urged the Mills students to get the rules changed so that they could become test pilots. James Webb sidestepped that sensitive point and said, "If a woman *scientist* has a need to go into space, we will shortly have safe transportation to bring her there and back."

The reporter who covered the symposium with admirable objectivity for the *Oakland Tribune* was Marion Dietrich, one of the twins who had reported to Albuquerque six years earlier to be tested as a potential "lady astronaut" at the Lovelace Foundation. She had made space her beat.

Two years later, in 1967, NASA announced it had selected a new group of astronauts (four other groups had been selected in 1962, 1963, 1965 and 1966). They were all scientist-astronauts who did not have to have been test pilots, and women had been invited to apply. Seventeen women scientists did, only to be turned down by the National Academy of Sciences, a clearinghouse for NASA. These applicants included several who had been the first women at their respective institutions to earn doctorates in science or engineering, and all were working as scientists on projects connected to space travel. Their frustrations saw print in *Parade* magazine in a November 1967 article by Jack Anderson, who, clearly outraged, suggested that the only way an American woman would make it into space would be as the guest of a

"Russian lady cosmonaut." Anderson itemized NASA's discrimination against women, noting that "NASA hasn't lifted a pencil to begin research on the space suits and other trappings a woman would need in space." Perhaps it had something to do with America's puritanical mores? Anderson wrote that the "gentlemen in charge of space travel apparently boggle at the thought of a man and a woman riding together in the close confines of a space capsule." He was probably right.

NASA insisted there had been no gender bias. But after interviewing eight of the rejected women and listing their qualifications, Anderson concluded that the women were at least as outstanding as the men who were selected. "Why," he asked, "would a normal American girl want to go soaring out of this world in the first place?" He had his answers from the "lady applicants [who] . . . share a love of science and a zest for adventure, [and] have the same dash and daring that made instant heroes of the original seven astronauts." Each had a different particular explanation. "We're going to find life on Mars," Gladys Philpott, a Ph.D. in histology and cytology, predicted, "and, oh boy! that would be the place to go."

Anderson was angry, but his anger did not travel. Becoming an astronaut was still too exotic an ambition to strike a chord with women trying desperately to get a foot in the door of less rarified places, like the Naval facility in Pensacola.

NASA did not seem to notice what was happening in the world beyond its rarified orbit. The momentum toward making women equal swept private education in 1968. In short order Harvard, Yale, Princeton and the California Institute of Technology admitted women, and Vassar accepted men. Women's studies entered the academic curriculum. At the same time, those resisting change spoke out. Unfortunately, some were in powerful positions, like Harry Hess, a Princeton professor who chaired the influential Space Science Board at the National Academy of Sciences. He declared that the possibility of space travel "looks bleak" for women because, among other things, there were the "millions of dollars" he estimated it would cost to design a suit. But "the very best reason for not choosing women astronauts," he stated, as if it was truly documented, "was the fact that women tend to get married. You spend all that money and time training them, and then they get married and have children. You can't expect them to leave a couple of kids at home and take off into space." Why not? Dads did. But this line of reasoning had kept women out of professional schools and would continue to, for a while. Of course mothers all over America routinely left their children, if not to go to paid jobs, then to care for other family members or to take on other urgent chores, but leaving kids behind was not part of womanhood's idealized image. And then there were women who did not have children, but they were simply omitted from the equation. One aim of the women's movement, and there were many, was to bring the image or women more in line with the reality in which most women lived.

Slowly the movement gathered momentum. On television the popular *Mary Tyler Moore Show* featured a woman who enjoyed male company and a career, but was not married. Feminist political action intensified and a slogan of the movement—"the personal is political"—took on another level of meaning with the marketing of the birth control pill and the Supreme Court's decision in the case of *Roe v. Wade* making legal abortion available. Sexual behavior changed, but not for everyone, and in an odd way, behavior changed before many people actually talked about sex.

Meanwhile, women who had joined the marches to defend Civil Rights and the marches to denounce American involvement in Vietnam began marching for their own cause. Women, who only yesterday, it seemed, had attributed the abrupt end of their career trajectories to personal inadequacies, now identified their problems as external to themselves. They called it a "glass ceiling," an invisible limit against which they could knock their heads but never penetrate. The women's movement did not win over every woman, but it drew a lot of attention and spurred an organized female opposition. Some of those who opposed feminist demands responded by organizing anti-abortion groups and lobbying to defend respect for women's traditional domestic role. But whatever a woman's political position might be, whether or not she considered herself politically engaged, pressure from the activists changed her life.

NASA, eventually, sensed the changes. Through the sixties its leaders had looked with blinders at the Moon, seeing themselves in charge of a great Space Race. Whatever scientific spin-offs might come from the Moon landings were always second to simply getting there first. NASA learned that the Soviets had lost their rocket genius, Korolev, in 1966, but they did not know how important he had been. Nor did they know that the Soviet space program had suffered a series of accidents and disappointments so severe that after sending a robotic vehicle to photograph the dark side of the moon, the Soviets naturally abandoned the the idea of piloted flight in 1968 and had decided to concentrate instead on building a reusable space station. The next year, 1969, saw the Apollo landing. The Moon landings commanded enormous attention for the first few times. But the public soon tired of watching men gather rocks and dust and exclaim about the view; people began to lose interest and NASA actually paid the networks to cover *Apollo 16.* NASA's risk assessors figured that it was statistically time for an accident and suggested stopping while they were ahead. NASA cancelled the final Apollo missions.

Meanwhile, in 1969, NASA had added another seven pilot-astronauts to its roster with transfers from the moribund Air Force program. This left NASA with a long list of astronauts waiting for a turn to work in space. They did not need new astronauts. The issue of adding women to the roster was temporarily shelved, even as pressure mounted from women's groups to do something.

Little girls began playing with Astronaut Barbies in 1965 when the Mattel Company released the shapely 10-inch plastic doll dressed in a silver space suit. The 1964 Barbie had been sold in the costume of an airline stewardess, with her boyfriend, Ken, dressed as a pilot. Now, while Gemini spacecraft were orbiting the Earth with only men inside, little girls fantasized about becoming astronauts themselves. For the first time Barbie, who had always been marketed as having a job, held a job that in the real world was open only to men, and she sold well.

In 1970, with men visiting the Moon, NASA looked to its future and the possibility of sending women into space. Charles Berry, then director of life sciences at the Johnson Space Center, mentioned a proposed mission to Mars, which would take much longer than the commute to the Moon, probably a few years there and back. He told the press that "it would be more comfortable if we took women along on long flights." More comfortable for whom? Although there were allusions to saving space for women on spacecraft, there was virtually no mention of allowing women into the astronaut corps as a matter of fairness. Which did not stop some astronauts from thinking about women. Michael Collins, the third member of the *Apollo 11* crew, had apparently been daydreaming while waiting for his colleagues Neil Armstrong and Buzz Aldrin to return from the Moon. He later wrote: "the possibilities of weightlessness are there for the ingenious to exploit. No need to carry bras into space, that's for sure. Imagine a spacecraft of the future, with a crew of a thousand ladies, off with Alpha Centauri, with two thousand breasts bobbing beautifully and quivering delightfully in response to their every weightless movement . . . and I am the commander of the craft, and it is Saturday morning and time for inspection, naturally. . . ."

Unique in a government agency, NASA had inherited the support but also the dreams of a science fiction constituency. A case can be made that fantasies from Jules Verne to Buck Rogers prepared the American imagination for space travel. The impetus to send spacecraft to the Moon may have been the space race, but the pressure on Congress to fund, and continue funding space programs, came from men and some women, too, enthralled with the dream of visiting the stars. This important constituency did not share NASA's chauvinism. They responded instead to writers like the prolific Isaac Asimov, who in February 1971 wrote in the *Ladies Home Journal:* "Only the moon can be explored in celibate austerity without undue strain. The only way for NASA to avoid permitting women in spaceships is to freeze the program at its present level or dismantle it altogether. Barring that, space exploration cannot remain an exclusively male preserve for much longer."

We know that by 1972 NASA was thinking seriously about accepting women as astronauts because starting that year its files have suggestions and drawings of possible toilet facilities for women. But the agency's incipient shift was kept secret. When Ena Naunton, an English emigrant in Florida and the only female reporter at a NASA press conference for the Aerospace Medical Association in 1972, was skeptical about the continued absence of female astronauts, no one answered her. Why, she wanted to know, nine years after Tereshkova's flight, was any mention that another woman join a space mission still met with lewd jokes and references to supplying male astronauts with sexual favors? She concluded: "One gets the feeling that women are to NASA what the camp followers were to military encampments of the Middle Ages. Creature comforts. Period."

NASA thoughts were elsewhere. They had cancelled the last planned Apollo missions and in 1972 found a way to use some of its leftover hardware. This became *Skylab,* the first American space station (the Soviet *Salyut 1* was put in orbit the year before), which began as a continuation of the Moon missions and was initially called the "Apollo Applications Program." It was a way for NASA to use some of its leftover Apollo astronauts in an orbiting workshop. *Skylab,* with 17 cubic feet divided into two levels, had a dining table, three tiny bedrooms for the three resident astronauts, a shower and bathroom. It offered more comfort and privacy than the not-yet-built shuttle ever would.

Skylab hosted three missions, two in 1973 and one in 1974, during which some astronauts spent as long as 84 days aloft. The missions revealed how a crew of three men could live and work during an extended tour in orbit. By the end of 1974 NASA knew that astronauts who live for a long time in microgravity suffer some physiological damage, most, but not all, of which disappears after the astronauts return home. They also discovered that while no astronaut had gone mad in space, several had become very angry. They became generally hostile to ground control and needed help managing their anger. In contrast to some Apollo astronauts for whom a few days on the Moon triggered religious piety, the *Skylab* crews remained spiritually unaltered.

Skylab ended in 1974 and there was still no sign of an American woman in space. *Skylab*'s putative purpose was to learn how the human body responded to long periods of weightlessness. However, with only male astronauts, the researchers studied only male physiology. From the perspectives of Mrs. James A. Lovell and Mrs. James A. McDivitt, wives of retired astronauts, that was as it should be. "I don't think they have ever found a woman capable of being an astronaut," Mrs. Lovell said. As far as she was concerned, women's response to weightlessness was not something she wanted to see explored.

Astronaut wives were then consulted about their husbands' working situation, as were the wives of firemen who faced the threat of female *firefighters* invading the twenty-four-hour camaraderie of the firehouse. This was a time of change and the struggle of some women to work alongside men in traditionally male space was met by the uncertainty of women married to the men whose work-space was about to be changed. Many of the women who had chosen marriage to men who risked their lives in the line of duty wanted society to leave their lives alone. They did not want to risk their husbands' marital commitment to women they saw as predatory. As hearth-warming partners of heroes, they felt they were defending a way of life. Neither Mrs. Lovell nor Mrs. McDivitt ever applied to become astronauts. Their identities were vested in the system that John Glenn had defended a decade earlier, a system in which the men went out to work, courting danger if they had to, while their women waited quietly for them to come home.

Skylab could have supported a woman. It was large enough to have provided plenty of privacy and there were women scientists up to the task. It would not have been beyond NASA's technical ability to provide them with toilets, but the proposal in 1973 was not cost-effective. NASA had already invested in too many male astronauts-in-waiting. Ena Naunton observed that NASA apparently wasn't interested in what would happen to women on long space trips, although these were assumed to be about to happen in the early 1970s. Most people, even at NASA, assumed that women would help colonize other planets and would be part of any extra-orbital exploration and would have to be factored in eventually. However, Charles Berry agreed with Naunton's observation. NASA was interested only in men: "We have no plans to send any women along. Medical science will examine a *man's* body in weightlessness." That is what *Skylab* was all about. He had talked casually about sending women on long missions, but *when* remained in some undetermined future.

The possibility of women astronauts posed problems that NASA seemed eager to avoid. Women menstruate. This subject, even the word, was seldom mentioned publicly or in mixed company in the early seventies. Among women the subject of their own bodies was a new topic of conversation that became a focus of the "consciousness raising" groups that flourished in the early days of the women's movement. At these grassroots meetings in living rooms all over the United States adult women shared their experiences with sexuality and their frustrations with life in a male-dominated society. They became familiar, with the help of the best-selling *Our Bodies, Ourselves* in 1971, with their own biology and anatomy. But what was no longer a mystery to them was still a mystery to many of NASA's physicians.

NASA's medical advisors worried, with reason, about the response of all bodily fluids freed from the tether of gravity. They feared that menstrual blood would not descend in microgravity, but might back up, perhaps causing aneurysms or at the least severe internal bleeding. There had been similar fears about the urinary tract, male as well as female. These had proved baseless to the men in space, but female physiology had yet to be tested and it was impossible to simulate weightlessness on Earth. By the end of *Skylab,* many of the apparent physiological phenomena of weightlessness had been documented, including the troubling effects on bones, kidneys and the inner ear. Whether women's bodies would respond the same way as men's, or how weightlessness would affect the female reproductive system, were still questions without answers. Indeed, no one at NASA had even asked the questions.

Outside of NASA's peculiar culture, Americans in the seventies were comfortable with the idea of sexually and racially mixed crews on spacecraft. The television series *Star Trek* attracted viewers in almost every age and social group. Even those who were not fans knew that "Space [was] the final frontier." Viewers followed the adventures of the Starship *Enterprise* as it explored strange new worlds, "To boldly go where no *man* had gone before." Although its first run lasted only three seasons, ending in June of 1969 when the virgin Moon was still innocent of human footprints, *Star Trek* had captured the American imagination. Its twenty-third-century heroes, Captain Kirk, Mr. Spock and the female black Lieutenant Uhura, traveled between galaxies coaxing the inhabitants of untoward planets into compliance with the peaceful, mature and egalitarian ways of Earth.

In his first draft, *Star Trek*'s creator Gene Roddenberry had made the captain female. The series' sponsors weren't ready for such a giant leap, so Roddenberry rewrote Captain Kirk as a man. As captain, Kirk spread 1960s American values throughout the galaxies. But when a reporter for the Soviet newspaper *Pravda* (who had obviously been following the program) complained that the crew should include a Russian, because of Gagarin's flight, Roddenberry obliged. In its second year on the air Ensign Chekov joined the crew. *Star Trek* was a surrogate for American liberal futurists. Breaching the walls of a society still struggling to provide civil rights for African-Americans, in one episode Captain Kirk kissed Lieutenant Uhura. This was television's first interracial kiss. *Star Trek* described a future world where female astronauts wore 1966 miniskirts and everyone, human and alien, worried about the late-twentieth-century problems of race and imperialism.

In the polarized political culture of the sixties and seventies, *Star Trek* and its growing band of "trekkies" were an unsought but generally welcome gift to NASA. At a time when NASA could offer the public only three *Skylab* missions, trekkies kept enthusiasm for space travel alive. *Star Trek* preceded the Moon landings, and its reruns, followed eventually by feature films and spin-off series, continued long after Apollo was history. NASA administrators saw that some of its most loyal supporters were aficionados of the *Enterprise. Star Trek* proved that interest in space travel had not died out. It had simply found another outlet, one where sociopolitical realities did not impede the imagination.

Of course the reality of the space program was light-years away from *Star Trek.* NASA was mired in the politics and personal ambitions of America's leaders. The Nixon administration looked on the Apollo program as a Kennedy legacy, despite the fact that NASA had been created by Eisenhower, whom Nixon had served as vice president. It was probably more about the budget and Vietnam that made Nixon let his Office of Management and Budget cut the allocations to NASA; and even with those cuts he had to face a Democratic Congress eager to prune the space budget even further to pay for domestic programs. According to Joan Hoff, "Nixon had no choice but to opt for *some* kind of human spaceflight project to succeed Apollo; the astronauts provided the necessary human element of a science that was largely unintelligible to the average person. No president in the 1970s wanted to be responsible for killing the only compassionate symbol of space exploration: the astronaut in orbit." Without a financial commitment, the design and construction of the space station and space shuttle suffered delay after delay.

NASA's long-term plan called for an orbiting station and a reusable transportation system to ferry people between that station and Earth. In 1971, it had to choose between going ahead with either the shuttle or the space station. NASA opted to start with the shuttle. In 1972 Congress agreed on funding. NASA predicted that the shuttle would be flying by 1978 and the space station, which the shuttle was supposed to serve, was to have been operational by 1986. These deadlines would be postponed frequently during the next decades.

But throughout these years loyal trekkies lobbied to keep the shuttle program alive and organized a letter-writing campaign that succeeded in getting the first orbital shuttle named after the *Enterprise. Star Trek,* a cinematic fairy tale, possessed the glamour and excitement that NASA, a government agency, did not. Locked into annual budgets and obliged to please hundreds of masters, NASA had to avoid anything contentious.

Since 1965, NASA had always struggled to get money for its programs and often ran at a deficit. There was a constant effort to win public support. It could not have pleased the agency when in 1973 *MS* magazine, then the organ of the National Organization for Women, ran a long article on the feminist issue that would not die—the saga of the thirteen women who had passed through the Lovelace Foundation. The author described the experiences of Jerrie Cobb and the FLATS as an ordeal of misunderstood promises and false expectations about the possibility of a woman becoming an astronaut.

The agency bowed to the inevitable and announced in 1973 that it would soon look for women astronauts. The *San Francisco Chronicle* noted that intention a year later, writing that NASA had finally "acknowledged women's lib." It would soon be selecting a new kind of astronaut for the new "shuttleships"—payload specialists. These astronauts would expertly handle the payloads—be they scientific experiments or the launching of special satellites—but they were not pilots. Some of these new astronauts would be female. Bob Parker, a scientist-astronaut selected in 1967, recalled in 1999 what his colleagues said about women in 1974: "After *Skylab,* it was generally accepted [that] women would have a role in the shuttle and the station. . . . The lowliest people and the people from headquarters said we would be flying women because we would need people to cook the meals and basically run the station."

But in 1974 Parker spoke differently to the press: "We're not talking about dizzy blonde secretaries, but reputable women scientists. NASA would not select Barbarella-like dumb babes, but smart women," by inference plain, but with credentials. Whatever they looked like, women could now fly with men because, as the *Chronicle* reported, the shuttleships would have "separate toilet facilities for men and women," and the women, like the men, would "have private living compartments."

The shuttle had won congressional approval in 1972. This was NASA's first major venture since Apollo and it was paired with plans to build a space station. The shuttle would be reusable and would ferry astronauts to the station, where they would do research, some of which would be contracted to private corporations whose fees would pay for the entire profit-making project. In the short term, however, which turned out to be much longer than anyone expected, the shuttle would serve as a laboratory and a launching platform for all kinds of satellites. It would carry between five and eight people at a time and have unisex sleep and work arrangements. To become operational, it needed toilets for women as well as men.

During the decades after World War II, Americans took pride in having the world's best plumbing. Travelers returning from abroad exchanged stories of the primitive toilet facilities in European and Asian countries, equating Western-style facilities with civilization. In the United States, this meant separate toilets for men and women. Designs for a female toilet, mentioned casually in press releases, fill many folders in the records of NASA administrators. A toilet for women led a list of seemingly insuperable logistical obstacles.

Toilets on spaceships have to be different from toilets on Earth. Mercury and Apollo astronauts simply urinated into a funnel they clamped to their penises, but something else was needed for women. The challenge of devising a system that would allow women to urinate in space was not as complex an engineering problem as the construction of the space station. But with all its cultural baggage, the problem loomed over the decision of when women could be invited to join the astronaut corps. As the budget continued to shrink, so did the size of the proposed shuttle. NASA designers knew that space constraints no longer allowed for private living quarters or separate "rest" rooms. They needed a single toilet with separately designed funnels so that women and men could use the same facility but with devices adaptable to their different anatomies. When they got to the drawing board, the design turned out not to cost very much at all.

Besides the places for navigating, eating, sleeping, showering and using the toilet, the shuttle included a 60-foot cargo bay adequate for carrying an instrument such as a satellite, a telescope and eventually parts of what would become the International Space Station. Eventually, of course, there would be a space station. But for twenty years the shuttle existed in limbo—a shuttle to nowhere that left Cape Canaveral, orbited the Earth for about a week, and then returned.

By 1974 NASA was on notice that the "woman question" would not go away. There was the idea, already floated, of a woman payload specialist, a scientist who would accompany an experiment, or a new category that was just taking shape, "mission specialist," a career astronaut who would fly many missions. In June NASA predicted that the space shuttle would fly in 1980 and for the first time said that it would carry female mission specialists as well as payload specialists.

But despite sporadic reports in the press, there was no mention of recruiting the much-discussed new mission specialists, a group that could include women. This left aspiring women in the early seventies with only one way to get into space, and that was to become a payload specialist. Quick to send one of their own, two of NASA's centers, Marshall in Alabama and Ames in California offered candidates.

Marshall had four candidates who had submitted experiments to be done on Spacelab, the laboratory module that had been scheduled for an early shuttle mission. But Spacelab was expensive and the project was stymied until NASA forged an agreement with ESA, the new European Space Agency, which promised to share both expenses and staff. The Marshall Space Flight Center was ready and released the news that one of four women scientists from its own staff had a good chance of being the first American woman in space. The Huntsville newspapers ran the story in 1977, announcing that "Four Wives" had been named candidates for Spacelab in 1980. Several months later the same papers announced that the candidates had been whittled down to one—Ann Whitaker, whom the *New York Times* identified as "a 38-year-old mother, NASCAR racer and NASA physicist from Huntsville, Ala., who could become the first American woman to fly in space."

Whitaker, chief of Marshall's physical sciences branch in the engineering physics division of the Materials and Processes Laboratory, was sent to Houston to go through the paces of orientation and preparation for space at the Johnson Space Center. She left her small daughter and husband, also a physicist, and gave up NASCAR racing for a while—but the family agreed that the promise of space travel was worth it.

At NASA's Ames Center near Palo Alto, California, another potential payload specialist, Patricia Cowings, had her name in the hopper. The local San Francisco newspaper, like the one in Huntsville, had a field day touting the probable selection of a Bay Area resident as the first American woman in space. Cowings was science director of the psychophysiology laboratory and had been eager to explore space since she began borrowing science fiction books from the New York public library near her home in the Bronx, New York. Like most of the women eligible to fly in the late seventies, Cowings was born just after World War II. Unlike the others, she grew up in an African-American family. New York City, which has always offered its bright and talented students a host of special schools, gave Cowings a place at the High School of Music and Art, where she studied and painted, and after school worked at the family grocery store, read a lot and watched television. Aside from books and her family, she says that the defining influence in her life was *Star Trek.* From the day she watched her first episode, she had space on the brain, and was determined, somehow, to get there.

There was no obvious path. At school in the midsixties she realized that "all the opportunities were for boys." White boys at that. "I concluded that, being a brown woman, I was at the bottom of the barrel."

But her father reminded her that she was a human being, "the best damn animal in the world, and that by virtue of being a human being I could be anything I wanted to be. And although he thought a woman's place was in the home, he thought his daughter's place could be in the laboratory, or wherever she wanted." She wanted, she realized, to be like one of her aunts who had a Ph.D. in psychology and had written a thesis on the IQs of African-Americans. As a student at the State University of New York at Stony Brook, Cowings, too, discovered psychology as the student of Neal Miller, the man who developed biofeedback.

From New York she went to graduate school at the University of California at Davis, and there saw an announcement of a course in space engineering. She did not have the prerequisites but approached the professor. "I told him he had to let me in because he had no women in the course, and who was going to do the shuttle's curtains?" She told him seriously that there were no life sciences students in the class, and so he let her in. The instructor was Hans Mark, the director of NASA Ames. He brought her to work at Ames that summer, and she has worked there ever since. At Ames she developed a technique and instruments for biofeedback to help astronauts monitor and control their autonomic responses—their pulse and heart rate. Then in 1978 she won a grant to try her system on the shuttle, and was selected as backup mission specialist for the project. Like Ann Whitaker she went to Houston to train. She recalls overhearing grumbling among her colleagues that NASA would never let women into the astronaut office. But in that matter they would be proved wrong.

For a while, it looked as if Cowings might be the first woman. The San Francisco newspapers asked her what clothes she would wear in space. Photojournalists photographed her in a mismatched collection that suggested that Houston had not yet come to grips with the fact that women were not just short men. Cowings enjoyed being described as perhaps the first woman astronaut, but she was never described as the first black woman astronaut. Being light-skinned, she could be mistaken for someone from the Mediterranean or Latin America.

Ames had become the center of "human factor" studies, an Air Force term describing the only component in an airplane that cannot be redesigned or refined—the crew. With the shuttle nearing completion and plans for a space station in the works, NASA needed to learn more about how the human body responded to space. They sought some kind of analog to weightlessness on Earth, and NASA arrived at the same conclusion the Soviets had—extended bed rest. The rationale was that in walking upright, humans had evolved the cardiovascular ability to keep blood evenly distributed despite the pull of gravity. In space and in bed, within a day or two, the body lost a lot of fluid and redistributed what remained. At the same time muscles tended to atrophy from disuse, and bones lost minerals. Ames had begun these bed rest tests in 1972 with a group of men. In 1978 they asked twelve female Air Force flight nurses to join a five-week experiment to see if women responded differently from men to bed rest, and theoretically to weightlessness. They wanted to measure specific physiological changes during simulated weightlessness, changes such as biorhythms, body chemistry and particularly hormones. Unlike the Lovelace candidates, these test subjects were not expected to be in prime physical condition, just healthy. Standards had changed since the Mercury, Gemini and Apollo days.

Whitaker and Cowings continued hoping, but their chances of being the first American woman in space receded as the shuttle kept being delayed. The press lost interest in both of them altogether when NASA announced in 1977 that the first American woman in space would not be a one-time payload specialist but a mission specialist, a career astronaut who was not a pilot. Trekkies took heart. Reality was catching up with their fantasy.

4

EXPLORERS OR PIONEERS

MARGARET "RHEA" SEDDON was already a heart surgeon in 1976 when she learned that for the first time in ten years NASA was recruiting a new class of astronauts—this time identifying itself as "an Equal Opportunity Employer." The agency had changed, as Seddon had always hoped it would, and now welcomed applications from women. NASA had protected its brotherhood through the Apollo years, but it had given in to pressure from women like Jerrie Cobb and those who grew up playing with Barbie astronauts and watching *Star Trek.* Women were at last invited into the great adventure.

Although NASA had announced in 1973 that an American woman would go into orbit, it had taken three years to figure out how the agency could manage without compromising the test pilot requirement. The shuttle provided a solution. Without tampering with the requirements for pilot astronauts—a job still reserved for military test pilots—they could now choose whether to take women as payload specialists or mission specialists. Payload specialists, like the scientists at the Marshall and Ames Centers, could fly as highly trained experts for a single mission. Mission specialists, on the other hand, would become career astronauts, anticipating a variety of responsibilities that they could handle because they would have advanced degrees in science, engineering or medicine. NASA chose the latter, mission specialists, for its first women astronauts, and said as much on the recruitment posters it distributed in 1976. The next astronaut class, thirty-five men and women selected the following year, would be the class of 1978.

Seddon, confident that NASA would eventually open its arms to women, was prepared when the call came. Long before she had entered medical school, before NASA had expanded its selection pool, she had decided to become an astronaut. Toward that end she had learned to fly, because all previous astronauts had been pilots, and had mastered surgery, a valuable skill wherever there are people, and a good fallback career should she not make it into the corps.

But she did make it into the corps in 1978. NASA acknowledged an end to the era of "good old boys," the men who had orbited in the sixties and seventies and proved that space, in the narrow sense, was a place where humans could survive beyond Earth's gravitational field. They also proved that American men could land on the Moon and bring back Moon rocks, that American men could survive the unknown in acute discomfort. Those astronauts, the Mercury, Gemini and Apollo astronauts, had been brave explorers. Now, NASA decreed, that stage was over and the next stage, the exploitation of space, was about to begin.

Who follows explorers? Why, pioneers.

A NASA spokesman spelled out this vision in 1982: "Now we're sending up settlers. The shuttle is the new Conestoga wagon." He was explaining that settling space was like settling the American West. That having explored the territory, it was time to settle it, which explained why NASA now needed women. Women, everyone understood, made houses into homes.

This sequence of events may have made sense on the American frontier; it did not ring true in space. Space is not a prairie or a wilderness. The countless stars are not subject to cultivation. From the perspective of the individual astronaut, there is literally *nothing* in space. The shuttle was designed as a taxi to ferry astronauts between Earth and a space station, a destination whose design would undergo countless revisions before it was completed, and whose purpose, even as it was complete enough to house occupants in the early twenty-first century, was not totally clear. During the shuttle's first two decades, as plans for the space station were frequently revised, the shuttle, officially designated a Space Transportation System (STS), served as an orbiting laboratory for commercial and government research, and as a launch pad for cargo that included space telescopes and military and commercial satellites.

The explorer-pioneer metaphor also did not work because female astronauts were not homemakers, and neither were the male mission specialists they flew with. The shuttle was not a home, and life inside did not reflect any earthbound habitat. Moreover, the tasks of the astronauts were not gender-specific. NASA spokesmen described life on the shuttle as normal, as just another kind of scientific laboratory. The public relations office emphasized how different it was for an astronaut in a shuttle from, say, life in an Apollo spacecraft. The shuttle made spaceflight "more a routine than an adventure." The new space story was "about people in space, not super humans."

That made good copy, but it wasn't true. Of course these astronauts were not super humans, but neither were John Glenn or Neil Armstrong. They were, however, among the first humans to live in an air pocket in the void of space, and when they went, they were not following in anyone's footsteps. Rhea Seddon had no role model because there were no female models for American women astronauts. It was up to her to become a model for the women who would follow. When these women joined NASA, they came as explorers under the superficial guise of pioneers.

The mystery of space burned in Seddon's soul. She was ready to do whatever NASA needed done, routine or adventure, old vision or new. She had been preparing for space travel ever since she had seen *Sputnik* in the sky over Murfreesboro, Tennessee, when she was eleven years old. The Soviet satellite had convinced the administrators of her small Catholic school, along with school boards around the country, that it was their patriotic duty to improve their science and mathematics curricula. Accordingly, the school hired a new science teacher in 1958. Among the teacher's demands were science posters from each student. Seddon recalls her seventh-grade entry, in which she conjectured about what might happen to humans in space: "I can remember drawing that little space man and talking about G forces and the vacuum of space and that you'd have to take your environment with you. And years later I realized that I had gotten interested back in seventh grade."

Science was making headlines in the fifties and children were reading them. The next year Seddon's eighth-grade science entry poster explained the double helix. Both space and genetics had caught her interest and they remained her passions. A daughter of a Yankee, Harvard-educated lawyer father and a mother who was a member of both the Daughters of the Confederacy and the Daughters of the American Revolution, Seddon learned proper southern manners, which she took along to the University of California at Berkeley in 1967. When she arrived as a premed the free-speech movement was sweeping the campus. Heedless of the political storms around her, Seddon joined a sorority, diligently went to classes and did her homework. "I think of myself as Forrest Gump, sort of going through life and all of these peripheral things are happening—the free-speech movement and the war protests—and I went about going to my physiology classes and going past the [protest] groups on my way to chemistry lab." Through all the turmoil, Seddon studied hard, waiting until her senior year to take a break from the laboratory to do what she'd dreamed of doing for years—learn to fly. She figured that by the time NASA took women astronauts, they would also need physicians.

Seddon had always made decisions quietly, ignoring advice from the gallery. When she told people at home in Tennessee that she was interested in scientific research or medicine, they had dismissed her with a "Give me a break. Women don't do that." Rather than respond, she simply left the South. But even at Berkeley, where she did not find gender discrimination or discouragement from professors, she noticed that there were only a few women in her courses when she began, and even fewer as she approached graduation. When she returned to Tennessee for medical school, she was one of six women in a class of 115, and when she reached her surgical residency, she was the only one. On asking to use the doctors lounge, she was told "no," that "the guys wander around in their underwear. They don't really want women there." She transferred for the rest of her residency. She never joined the woman's movement but knows that "there are a lot of women that I owe stuff to, folks like Wally Funk [a FLAT] who pressed NASA at the beginning."

In the spring of 1977 Seddon sent her application to NASA. At the time she did not know exactly what NASA was looking for, or that NASA had not expected so many physicians to apply. Most likely NASA had not yet decided what training and personality traits to look for in this new kind of astronaut. What NASA did know, from the public opinion polls it frequently commissioned, was that public interest and support for the space program had dropped steadily since the first Moon landings, and in 1976 its pollsters reported that women in general did not support space exploration. This should not have been a surprise.

Three years earlier in October 1973, Ruth Bates Harris was serving as deputy assistant administrator for equal opportunity, and was NASA's highest-ranking female employee when she submitted a report that had definitely *not* been requested by the NASA administrator James Fletcher. The report noted: "There have been three females sent into space by NASA, two are Arabella and Anita, both spiders, and the other is Miss Baker, a monkey." It went on to itemize how few women and minority employees NASA had, far fewer than all other federal agencies. Moreover those on the NASA payroll were clustered at the bottom end of the pecking order. Only 1 percent of its engineers were female. Fletcher, distressed by the unrequested report, fired Harris. He attributed his decision to her inability to "get along with her co-workers." This action placed him directly in the firing line.

The firing of Harris and her report were exposed in detail by Constance Holden in *Science,* the organ of the American Association for the Advancement of Science. Soon afterward, on December 4, Representative Charles Rangel of New York entered the *Science* article, *in toto,* into the *Congressional Record.* The report itemized NASA's failures to implement EEOC guidelines at the Johnson Space Center as well as at the Washington headquarters. And when Fletcher suggested that Harris, an African-American, had used the same tactics as the then infamous African-American activist Angela Davis, a host of members of Congress, including Senator William Proxmire, testified to her gentle demeanor. The report had accused NASA of being "insulated from the real world [as it is now]." She had added, as if to underline her point: "It is important to note that Russia has had a woman as a Cosmonaut since the inception of their space program."

The *Science* article concluded with criticism of "the overwhelming white male domination of NASA . . . an anomaly among government agencies." It quoted a female consultant to NASA who said that she never "walks down a hall with a bunch of papers because they'll think I'm going to the Xerox machine."

The storm continued through June 1974. By this time Senator Proxmire had held hearings before the Senate Appropriations subcommittee and ordered NASA's equal employment budget doubled. Later that year Harris was rehired to the applause of feminist organizations that had supported her case.

Harris was in NASA's Washington headquarters when Chris Kraft headed the Johnson Space Center in Houston, the center of the astronaut program. Kraft was no friend to the cause of women in *his* jurisdiction, and in a memoir published in 2002 he blames his woes on Lovelace: "Lovelace caused us more problems when he allowed some female volunteers to go through tests without our concurrence. One of them was the wife of a U.S. senator, and before long we were being dragged through the 'Why no women astronauts' controversy." After almost thirty years the subject still rankled him. He noted that in the seventies "the subject of including women never came up until it was raised by outsiders." Exactly who did he mean by "outsiders"? Women? Activists? Outside of what? His NASA? The women who wanted to be astronauts were all mainstream Americans. From trekkies to university students to physicians, they represented a broad spectrum of citizens whose ideas of patriotism and opportunity did not jibe with Kraft's or the clique of associates he had worked with since joining NASA.

Whatever reasons NASA offered for not considering women as astronauts in the past did not hold in the shuttle era. The special training that astronauts needed as mission specialists—learning to work in pressure suits, to bail out of an airplane and survive in open ocean, to use ejection seats from jet aircraft and to breathe oxygen—training that was common to all military pilots, was new to all mission specialists, male and female alike. As Carolyn Huntoon, then chief of NASA's biomedical laboratories at the Johnson Space Center, noted with an implicit sigh of exasperation: "No one could find any reason women could not fly."

So NASA stopped looking for excuses and took the plunge. After all the lobbying and agitating, NASA sent out its June 1976 announcement inviting women to apply to the astronaut corps. Some did, but not many. NASA had expected an avalanche of mail, but after six months, very few women had sent in applications. It looked as if the public, especially women, had lost interest in the space program, or deemed it too expensive and too peripheral to problems that needed immediate attention, like poverty or the social upheaval left over from the Vietnam War. Women were especially disdainful. Although there were more women scientists and engineers in the United States in 1976 than there had been in 1965 when the students at Mills College had heard about the space program at their Founder's Day ceremony, the pool of qualified women interested in NASA wasn't very big. After six months only ninety-three women had sent in applications. It was embarrassing. Headquarters asked Kraft to develop plans to recruit women and minorities.

Kraft began by creating an Astronaut Selection Board charged with developing new processes for evaluating applicants, and he appointed several astronauts to this board, including Bob Parker. He also appointed Joseph D. Atkinson, an African-American. For matters of gender, he appointed Carolyn Huntoon. Her first suggestion was that no one ask women recruits any questions that were not also put to men, especially questions concerning their marital or family plans—a personnel tactic under attack by feminists nationwide.

The new board then sent out thousands of additional bulletins, but this time targeted colleges, universities and aerospace companies, as well as the personnel officers at all NASA centers—anywhere female scientists and engineers worked. And that was not all. Additional letters went to the Association of Women in Science, the Society of Women Engineers and the American Association of University Women.

The new selection board had two missions: to get women to apply to the corps and to find minority astronauts, men as well as women. Newly sensitive to the way others saw them, NASA's administrators began to notice how, wherever astronauts made public appearances, someone would ask when nonwhites and women would be going into space. Now that NASA wanted them, the candidates seemed to have vanished. The board then contacted

Nichelle Nichols, the African-American actor who had played Lieutenant Uhura, the communications officer on the Federation Starship *Enterprise* on the television series *Star Trek.* The series was enjoying a post-television afterlife throughout the seventies at trekkie conventions, where the cast made regular appearances.

It was at one of these conventions in 1975 that Nichols listened to a NASA presentation and was converted, she writes, to the cause of real space travel. She began to think of the character she had played as the fictitious grandchild of currently living NASA astronauts and this made her uneasy, because, she recalls thinking: "Where the hell is 'me'? There was no one in the astronaut corps who looked anything like me. There were no women, no Blacks, no Latinos. I could not reconcile the term 'US Space program' with an endeavor that did not involve anyone except white males. . . . Why weren't there women and minority astronauts?'" Nichols didn't let NASA's policies dim her enthusiasm and on her own initiative she visited every NASA center. She became a familiar face, so when NASA was asked why its recruitment efforts were inadequate, the agency asked for her help, awarding $49,900 to her production company, Women in Motion, to publicize the search.

She loved the assignment. "When we start colonizing space I don't want to see Blacks being the chauffeurs and tap dancers and women taking jobs as secretaries and maids." She knew the statistics, how few American women actually studied engineering, math and science, but she also knew that there were plenty of qualified women out there. It was her mission to find them and get them to apply. Six months later, whether it was Nichols' efforts or simply momentum, NASA had collected another 10,000 applicants, including 1,500 women.

In 1977 the agency conducted two overlapping searches, one for pilot astronauts, men a lot like the old elite, and a second for the new mission specialists, some of whom would be military men or women who were scientists and engineers, and others who were civilian scientists and engineers.

The selection board knew that they wanted pilot astronauts with the same experience and motivation that characterized the earlier astronauts. But they also knew that they did not want the kind of people who would soon be described in a best-selling history of the Mercury program as having "the right stuff," which meant, in several instances, cowboy-style daredevils who had succumbed to the limelight. The board wanted people with skill and daring who were also team players (no one would be asked to fly alone again). This time around the selection process would eliminate men who either did not respect women or who respected them socially but did not want to work with them. This turned out to be largely a nonissue because America's universities were changing rapidly and by 1977 had graduated plenty of talented and skilled men who were used to female colleagues and had no problem working with them.

The selection board knew how to eliminate what they didn't want, but they still had to decide what they *did* want. The fact of the matter was that there was no job like it on Earth, and not knowing exactly what challenges lay ahead in space, intuition played a significant part in the selection process. The board needed candidates who were experts as doctors or scientists or engineers because they would probably be asked to conduct medical or scientific experiments or solve technical problems either in the shuttle or as part of an experiment. But how technical or scientific these problems might be was hard to foretell, which is why they needed people trained as experts who could transfer their expertise, if needed, to another domain. They needed people who could shift focus without losing the edge that had made them desirable in the first place—team players who could be fielded as shortstops but could, in a pinch, be relied on to pitch.

Since the Mercury Seven's selection in 1959, the application process had changed along with almost everything else in NASA. Candidates still had to be in good shape, but they no longer had to swallow three feet of rubber tubing or endure most of the other trials that the Mercury men and the FLATS had endured at the Lovelace Clinic. By the end of the summer of 1977 the applicant pool had been winnowed down to 208 finalists, including twenty-one women. They were all invited to visit Houston for a final set of tests.

One of the twenty-one was Sally Kristen Ride. She was twenty-six years old when she arrived, a graduate student in astrophysics at Stanford who had seen the NASA bulletin in a student newspaper. She had had no specific plan for her future but, as a science fiction fan, felt that the chance to do physics in space was too good to pass by. So she filled out the forms. When invited for an interview, she embarked with enthusiasm mixed with vague trepidation. She remembered hearing how the Mercury astronauts had been spun in centrifuges and dunked in ice water. She found instead an ordinary medical examination with stress tests on a treadmill, and learned that it was all right to wear contact lenses to achieve "perfect" vision. Candidates had to be physically fit, but they did not have to run ten miles a day or do a hundred sit-ups. NASA would take care of their fitness later in the gym and through water-skiing and dropping them with parachutes into the ocean.

Ride would probably have done all right on any of the earlier tests. Ranked eighteenth nationally as a high school tennis player, she had transferred to Stanford after a year at Swarthmore College in Pennsylvania, partly to return to California, and partly for Stanford's tennis team. Tennis had won her a scholarship to Westlake, then a girl's preparatory school in Los Angeles, where a formidable teacher had encouraged her to be a scientist. Somewhere along the way she joined NOW, and as she said at the time, and again recently, she "owed her career to the feminist movement."

In an interview in *MS* magazine in January 1983, Ride described part of the Houston tests. "I'd never seen a psychiatrist before," she said, but she recognized a "good cop, bad cop" mental health examination setup when she met it. "The first guy was very warm. 'Tell me about yourself. Do you love your mother? Why do you love her? How does your sister feel about you? The second psychiatrist snapped orders like, 'five-seven-four-one-three. Repeat that backward!'" Ride could not have known that these psychological tests were constructed strictly to weed out people with serious mental problems.

But NASA's psychological tests did not "select in." They were not designed to find people who could get along in a small isolated group like a shuttle crew for long periods of time because, although the shuttle was small, with about as much space as a large camper, NASA did not envision trips longer than ten days. NASA was not especially interested in personal psychology or group dynamics at this time, although when they interviewed potential astronauts they looked for some assurance that the candidate would not crack under the strain of enforced isolation, sensory deprivation or the strangeness of space life. All they used were standard tests for mental health.

But candidates had to be able to get along with each other and with the literally hundreds of people in Houston and Florida who collaborated in sending a crew into space. Terry McGuire, a psychiatrist who consulted with NASA for the 1978 selection, explained that all of the women chosen had taken the road less traveled. These women had always made their own decisions and probably listened less than the average person to what other people thought. As scientists, engineers or physicians, they had studied alongside boys and men all of their lives and were at ease in male company. And unlike a great many American women in those days, none of the applicants revealed any fear of success. At the end of the interview process each candidate met for an hour-long, ten-on-one, face-to-face meeting with the entire selection committee. This included Apollo astronaut John Young, one of the men who had walked on the Moon, and George W. Abbey, already a powerful NASA figure who had been with the agency since 1964, when he had been assigned by the Air Force to work with the astronauts in the Apollo program. Abbey had left the Air Force in 1967 after playing a key role in the investigation of the oxygen-fed fire in an *Apollo* spacecraft that killed astronauts Roger Chafee, Ed White and Virgil Grissom. He would devote his career to NASA and to the astronaut corps.

Abbey remained a civilian with NASA, and with the exception of six years in headquarters in Washington between 1988 to 1994, he worked with the astronauts in Houston. A secretive man, he is reputed to have had a hand in almost every important decision dealing with the astronaut program and he had a strong, probably decisive voice in astronaut selection. His figure would loom large in the life of every astronaut. Always referred to as Mr. Abbey, his official title would change over the years, but one thing remained constant: his crucial role in deciding who would be accepted into the astronaut corps, and once in the corps, who would fly.

It was George Abbey who chaired the selection board in 1977, and it was he who defined the criteria for the new mission specialists. They didn't have to be great scientists because the science they were going to be doing was really someone else's experiment. If NASA needed someone with a highly specialized background in one scientific field, he explained, they had another special category, payload specialist. They could fly the specialist for that one flight. NASA needed these highly specialized people for the all-science flights. But for career astronauts, they were looking for people who would be asked to do one thing on one shuttle and something totally different on another, people who had experience working on a team.

According to Abbey, NASA wasn't affected by either the women's movement or public opinion or, for that matter, the results of its frequent opinion polls. Women were added to the mix, he says, "because at that point in time, they met all the criteria." But women had met all the criteria for scientist-astronauts in 1967, when none were accepted. Abbey's views on the women's movement notwithstanding, nothing had changed since the last call except the women's movement. Abbey had seen that it was time. Tommy O'Toole, then the space science reporter for the *Washington Post,* agrees, saying the selection board was very sensitive to public opinion. He recalls that, initially, the board selected only two women, but they were asked to take a second look at the list and try again. They were told that two was "not enough." And it is likely that Abbey told the board that, having finally committed NASA to women astronauts, it was important that they not take a token two but a large enough group to have an impact on the all-male corps, and on the world that was watching.

In selecting the Mercury Seven in 1959, NASA had decided to take men accustomed to facing extreme risks—test pilots. But as one of the doctors running the 1959 selection recalls: "We had one [other] thought: they would then be a homogenous team, including their wives." NASA had the idea that these brave, white, middle-class families were the kind that all Americans could identify with. In 1978 homogeneity was precisely what NASA wanted to avoid.

NASA also avoided mentioning the earlier criterion of bravery because that suggested that traveling in space was dangerous. The shuttle program was supposed to demonstrate that space travel had been somewhat domesticated, each mission a little house on a new kind of prairie where worker-astronauts led ordinary work lives. NASA presented this picture to the world at the press conference in Houston in January 1978 where, for the first time, all thirty-five members of the class of 1978 faced the public. In addition to the women (all white), the class included the first African-American and Asian-American men.

The women shared skin color and graduate educations—there were two physicians, two scientists and two engineers—but otherwise they came from different backgrounds and approached the world differently. Physically, all had passed the same medical exam as their male colleagues, with a few special exceptions. For one thing, pregnancy was unacceptable, at least during spaceflight, and a history of endometriosis was also a disqualifying factor, based on the fear—later discredited—that zero gravity or excessive radiation would aggravate the condition.

Before selection, NASA's life science people barely mentioned menstrual problems, but afterward the women discussed the issue themselves. Seddon recalls that there was some concern about whether women would be incapacitated with cramps around the time of their periods. "The way we looked at it," she said, was that women who "had gotten to this point in their careers had either not had major problems or had learned to deal with them so they [did not have] an impact on their careers. So they probably wouldn't have one on space flights." The women always had the option of taking hormones—birth control pills—to prevent their periods from occurring during flight.

In the seventies, NASA nursed dark fears about menstruation. Seddon said that they worried about retrograde menstrual blood moving up into the fallopian tubes and into the abdominal cavity, which would have to be treated by suctioning the blood away. From what she knew, Seddon said, "It just didn't ring true. Women have had menstrual periods during bed rest and in a variety of situations. I think we just persuaded NASA to consider it a non-problem." Then somewhere along the line, Seddon said, one woman had a menstrual period in space and "it came on time, was the same as her usual [period].. . . It was one of those things we didn't worry about."

What did concern them was the calendar. Scheduled to begin flying in 1980, the shuttle was postponed again and again. The new astronauts spent their first year as ASCAN (AStronaut CANdidates), experiencing weightlessness as they were flown in the T-38, a jet that flies in parabolic loops leaving passengers weightless for about 20 seconds at the top of each arc. Few liked it. What most of the astronauts did like was flying. Those who didn't know how, learned, and some, like Sally Ride, took to the skies so easily that they earned a pilot's license. During that candidate year they also learned to parachute onto land and sea, and they mastered the intricate control system of the shuttle. The Ph.D.s found it a lot like being back in graduate school. The doctors found it somewhat easier than their internships and residencies.

The routine applied to women and men alike, but the women in the class of 1978 had some extra duties. It was left to them to decide what, if any, new arrangements would be needed for women in orbit. Since their arrival they had willy-nilly become a community. As women in the first new class in a decade, they were a novelty. After all the speculation about the first American woman in space, the press finally knew who *she* was. *She* was one of the six already living in Texas. Because the press didn't know which one, they went after them all.

Since the Mercury Seven, NASA's public relations office had cheerfully supplied an eager press with astronaut stories. The public may have lost interest in how the spacecraft would be paid for, but they had not lost interest in the lives of the people who got inside those vehicles and blasted into space. NASA channeled the press into organized conferences but photographers and reporters still lay in wait whenever the women left the privacy of the center. The pressure of the spotlight was intense, and although none of them liked it, they did not circle their wagons for protection.

That was a relief. The astronaut office did not want a subgroup forming inside the corps. When they were not working out or practicing on simulators, all astronauts worked at desks in offices that they shared with other astronauts. The women in 1978 were assigned offices with male astronauts, both rookies and seasoned fliers, but not with each other.

Like the men in their class, the six women—Rhea Seddon, Sally Ride, Kathy Sullivan, Anna Fisher, Shannon Lucid and Judy Resnik—were smart, well educated and competitive. Unlike the men, they were all civilians. They came from both coasts and the nation's midsection and brought a variety of experience. Anna Fisher, a physician from Los Angeles, was twenty-nine and arrived with a new husband, who would be accepted as an astronaut in the next class two years later. Shannon Lucid, a chemical engineer from Oklahoma and a devout Nazarene, came with her husband and three children. Sally Ride, a generic Protestant, arrived with a new Ph.D. in astrophysics; Rhea Seddon, who was twenty-nine, arrived as a heart surgeon; and Kathryn

Sullivan, twenty-six, was a Ph.D. in geology who had toyed with pursuing undersea exploration. Judith Resnik, with a Ph.D. in electrical engineering, was a no-longer practicing but once orthodox Jew and a concert-level pianist. Every woman in this first group was an oldest daughter with supportive parents. None was from a wealthy family, but all had degrees from prestigious universities. Two were married when they joined NASA, and one of them, Anna Fisher married a physician who became an astronaut in the class of 1980. Before the first shuttle flight in 1982, two of the original six, Rhea Seddon and Sally Ride, had married fellow astronauts from the class of 1978.

All six could recall a specific childhood memory of a space milestone. Seddon remembered *Sputnik* followed by an infusion of funds into her school's science curriculum. Fisher, in school in San Pedro, in southern California, recalls listening to a transistor radio a classmate had brought to school as Al Shepard's flight was broadcast in 1961. And when sometime in the ninth or tenth grade someone asked what she wanted to do when she grew up, she said, "Well, what I'd really like to do more than anything is to be an astronaut and go into space and explore."

A few remember when gender became an obstacle to their ambitions. For Fisher it happened in 1967 when she applied to be a Senate page, and "discovered that they only took boys." This encounter with institutionalized chauvinism indicated to her that her secret ambition to become an astronaut might also be blocked. At the time "all astronauts were pilots, and all pilots were men." She paused and added: "I was already starting to think out of the box, but not *that* far out of the box." However, that one setback didn't send her into the women's movement, which to her was mixed up with other causes, like the anti-war movement. In fact, she felt lucky. "I feel like I probably came along right at the crest of the wave. If I had wanted to do the same thing in the forties, that probably wouldn't have happened. So I am very appreciative of the movement in terms of really pounding the way and opening the doors. But I wouldn't say I was really a part of that. I was so busy, in all honesty, getting my technical skills and studying."

Shannon Lucid, at thirty-three the oldest and tallest woman in the class, was born in Shanghai to missionary parents with whom, at the age of six weeks, she was sent briefly to a Japanese prisoner of war camp. Benefiting from a prisoner exchange with the Japanese government, she returned with her family to Oklahoma and the Texas panhandle, where she grew up. As a girl she developed a passion for aviation and chemistry, and earned a Ph.D. and a commercial pilot's license at about the same time. Neither of these helped her in 1964 when, wanting to be a pilot, she was turned down by commercial airlines and was forced to put her chemistry doctorate to use in Tulsa emptying bedpans in a nursing home.

Both of Sally Ride's parents taught, her father at Santa Monica City College and her mother, when her daughters were older, in public school special education. They thought of themselves as casually permissive parents in what they hoped was a nurturing way. Her mother recalls that she "mostly let them [Sally and her younger sister] explore." For her part Ride recalls that indeed they did. She can't remember a single time her parents told her not to do something she wanted to do.

Kathryn Sullivan, a geologist who had also grown up in Los Angeles, attended first grade with Ride at the same elementary school at the same time, but they didn't know each other. In the late 1940s her family had moved to Los Angeles, where her father, an engineer, had a job with the aviation company Pratt and Whitney. Sullivan lived abroad as an exchange student and discovered a gift for languages, becoming fluent in French and Norwegian. Insatiably curious, she was committed to going to the bottom of the sea when the unexpected opportunity to explore space offered her another venue.

The big question for her was, "We can look at the unknown, but can we go there? Can we figure this out? I wonder what we would find if we did [go there]? I wonder what living there would be like?" She relished the idea of doing something no one had done before. "Not because you want your name written down first, but because it's fun to actually be on blank, creative paper, being the first creative person."

Judith Resnik, an electrical engineer from Akron, Ohio, had parents who were devoted to her, but not to each other, and she seemed to bear the scars of their divorce. She was intensely private, but made friends easily, when she wanted to, and liked to cook and play the piano.

That was the group until 1980. Then NASA introduced another new class of astronauts to the press that included two more women: Mary Cleave, a civil engineer from New York, and Bonnie Dunbar, a mechanical/biological engineer from Washington. Both women had applied in 1977, been interviewed in Houston, missed the final cut, and had been encouraged to apply again. Cleave had returned to Utah, where she completed her doctoral dissertation, and Dunbar had taken a job at the Johnson Space Center as a guidance and navigation officer. Dunbar and Cleave, like Lucid and Fisher, recall frustrating collisions with institutionalized chauvinism.

They came from opposite coasts and very different families, but in both homes there were two working parents. In Great Neck, Long Island, Cleave's mother taught learning-disabled children and her father, a musician, commuted to New York City. Cleave was the oldest of three daughters; while her sisters talked about horses, she nursed a passion for flying. Aviation was an expensive, if not eccentric, obsession in the sixties, so to pay for flying lessons she worked out a fee-sharing arrangement with her parents. She took to the skies like a sparrow and had a license to teach passengers by her sixteenth birthday, two years before she could get a driver's license in New York. "My parents had to get me to the airport where I could legally take my clients up in the air." Scarcely five feet tall, she was turned down for a job as a flight attendant after high school. But small was fine in space and she relishes the irony of having been too small to serve coffee but big enough to ride into orbit.

On the west coast, Dunbar's mother worked alongside her husband on a farm they had homesteaded east of the Cascade Mountains in Washington, and there Dunbar shared farm chores with a sister and two brothers starting very early in the morning and, after school, continuing into the evening. She was working the night she saw *Sputnik.* "The Milky Way, the big stripe across the sky. Anything that was moving, you could see it because there was nothing else moving in the [sky in the] fifties." By the time she arrived in Houston, the night sky was laced with other artificial satellites.

These eight were the first American women trained as astronauts. But only one could be the first to fly.

The Johnson Space Center lies within the city limits of Houston, surrounded by the suburban community of Clear Lake, a series of residential subdivisions and malls that could be anywhere in the United States. At the turn of the twenty-first century, Clear Lake had shopping centers and motels at every price. Perpendicular to Interstate 10, which continues south to Galveston, NASA Road 1 leads past Space Center Houston, a for-profit tourist center, to the Johnson Space Center next door. A few old rockets stand like heralds on the drive to the Center offices. The rest of the Center looks like a college campus. Green lawns separate the buildings, which, as in most of the NASA centers around the country, are designated by numbers, not names. In the center of a "quad" near the nine-story administration building there is a pond and trees large enough to provide shade.

These women, like the men who preceded them, were beneficiaries twice over. They had profited from the *Sputnik*-spurred budget increases in science and math curricula, and they were helped by the women's movement. For while they stuck to their books, other politically active women pushed through gender enabling legislation. The first female astronauts walked through doorways that their activist sisters had pried open for them.

But they were unlike most of the women around them. Anna Fisher points out that before they reached Houston each one of them had all already succeeded in fields where women were a small minority. Fisher points out, "These were not your typical seventies women." They weren't used to being around a lot of other women. They were loner types who got along and found friendships outside the office. They were used to having men friends because they had lived in largely male worlds. They had all known what it is like to be the only woman, or one of a few women, in their field. Fisher recalled being the only woman in her study groups in medical school. What they were not used to was having female friends.

They had not needed role models but understood that they would become role models for future generations. They were the first, and people who do things first are not usually interested in following in someone else's steps. They did what they did, and still do, because it was possible. They liked being first in space. They had not fought at the barricades for women's issues because that was only their goal insofar as it could help them get into space. Only Sally Ride identified herself as a feminist. Endlessly in the spotlight, they grew weary hearing themselves called "the first women astronauts." They just wanted to fly. Judith Resnik declared that she wanted to be called an "astronaut," not a Jewish astronaut or a female astronaut. Resnik, alone of the six, did not live to have a second opinion twenty years later.

Did each of them want to be first? Kathy Sullivan said that is a question she never asked the others, but she "rather doubt[s] it. . . . In the sense of 'Do you want your name in the paper? Do you want to be famous?' I would suspect that was a scarcely present motivation—if at all." Yet she sees "that there was competition to be so good that you got the nod to be first and beat out the other guy."

Fisher remembers it a little differently. She can't see how any of them could have avoided wanting to be first. It was on everyone's mind. "How could it not have been? They were competitive people to begin with, exceedingly competitive or they would not have gotten where they were." Sally Ride is more circumspect: "Of course, we all wanted to be first, but we didn't know what the selection standards would be, so we just did the best we could."

But in spite of the tension, cooperation was the norm. Whichever woman would fly first would be in a crew; all astronauts are selected as team players and they were all team players. The intense schedule produced a kind of intimacy among all the new recruits, men and women alike. They worked together, relaxed at barbecues together, knowing that one of them would be famous as the first American woman in space, and insisting that it didn't matter who.

Under this pressure did the women bond as a group? Not exactly, because as Kathy Sullivan recalls, her closest friends were "probably Dick Scobee and Fred Gregory," men in the class of 1978. Yet she explains: "Certainly the six of us banded together." They had common interests, especially in the early days when they began to establish themselves. "We would make an effort to get together and talk about an issue, or talk about something that happened or if something had been proposed that was about women in the astronaut corps, we would pull together and get our views out. We would collectively pick a way to go."

As Seddon recalls, they were friends, but not *good* friends.

NASA used them all for publicity. Sullivan remembers that "there were times we'd all get thrown to the press." And if not the press, they were brought to events as what Seddon calls "potted palms," showcasing them to prove NASA's equal opportunity change of heart.

When the shuttle astronauts arrived in 1978, the roads were not paved and the streets around the Space Center had a frontier quality. The Mercury astronauts were gone but some of the Apollo and Gemini astronauts were still around and young girls trailed after them as if they were movie stars. The men in 1978 were greeted by camp followers like the young women who had met their predecessors. The women found a different welcome. There were no crowds of admiring men and, in fact, they suffered from what Mary Cleave called "PWS," "Professional Woman Syndrome," which is to say that men who were not astronauts saw them as unapproachable.

Some of the older astronauts from the earlier missions were friendly, Seddon recalls. "Others, I think, resented having women. They didn't think we'd do a good job and we were taking up a space of another guy. Most of their [these older guys'] relationships had been romantic. They didn't know how to treat women who were on the same level with men." The aura of being an astronaut that attracted female groupies to the guys did not work in reverse in Texas in the early eighties. They either frightened men or attracted publicity hounds. The women found it easiest to socialize with their fellow astronauts.

But not all of the men wanted to socialize with them. A few of the older astronauts did not want female colleagues. Al Bean, who had flown to the Moon and had come home to paint what he had seen there, told the *Texas Monthly* in 1978 that he and some of his buddies thought that women were being sent to do a man's job. He called spacecraft and computers "male things." He and his friends did not reject the women in their midst; they just felt it wouldn't work. John Young, who piloted the first shuttle flight, remembers arguments in the astronaut office about taking women. "They [the naysayers] just thought it was too risky." When asked whether it was the crew or the women who were at risk, he says with a grin, "Both." The guys were wrong, Young notes. "They discovered that women are pretty tough. They take care of themselves." Most of the older men adapted to the new demographic.

Kathy Sullivan remembers listening to the usual quips about needing women in an eventual space station, because someone would have to clean up. "I just tossed back the best line I could come up with . . . Only too happy to do windows, but only from the outside." That was before she did her spacewalk.

One of the responsibilities astronauts have always had is explaining the space program to the public, especially at schools and universities. NASA's public relations office fields the invitations and most astronauts accept those they can. In 1981, with eight women astronauts available, there were lots of invitations, but few from feminist groups. The struggle to get a place for themselves in space, hard fought and partially won, was not yet in the sight lines of organized feminists.

There were still women who did not want to see this particular change. The astronauts' wives were threatened by the presence of women who would work with, train with and fly with their husbands for months on end. Clear Lake had been their home for over a decade when the women astronauts arrived. It is not hard to imagine their feelings. A few must have even felt cheated. NASA had discouraged the first generation of astronaut wives from personal ambitions. When Patricia Collins, the wife of Apollo astronaut Michael Collins, was offered the opportunity to write a regular column in the local Clear Lake newspaper, NASA told her—ordered her—not to accept. These were military wives who had accepted a support role as part of the marriage pact. For them, the terms had not changed. Witnessing the arrival and integration of women who wanted the same careers as their husbands' and who wanted to work alongside their men as equals must have been devastating. Judith Resnik's father remembers her telling him that at first "the wives of the astronauts who were married kind of looked at her, wondering. But Judy kept her hands off. . . . And the women, the wives, accepted her."

Seddon recalls that the wives were gracious, but they did not at first understand that the female astronauts had the same busy work schedules as their husbands. She remembers a luncheon invitation on a weekday in downtown Houston, an hour's round trip from the Space Center. The astronauts could not leave work at noon and the luncheon never happened. It was a clash of women's roles, especially in the South, where women traditionally entertained in a certain way. Eventually the families of the astronauts met on weekends, and as the years passed, the divide disappeared as some astronauts married each other and most astronaut wives, like other American women, moved into the workforce.

While they waited to fly, the women drew assignments to prepare for those flights. Rhea Seddon, who had studied the effects of radiation on nutrition in cancer patients, was charged with the responsibility of finding a menu suitable for women in space. In a conversation with *Weight Watchers* magazine in May 1979, she spoke as an alumna of their program, having gained an unwanted ten pounds as a medical intern. Looking ahead to the flight, she explained that before the arrival of women, "each astronaut had a daily food allowance of 3,100 calories a day." But four of the women were under 5'5". That much food would have "turned us into blimps." So she had replaced puddings and cookies with fruits and vegetables. Weight gain, she worried, it turned out mistakenly, before she had been in space, could be an issue. "One of the problems in space is that you don't have gravity to work against, so you really have to work at it to burn calories."

Then there was the question of the women's hair. Cut it, tie it or let it float free? As with menstruation, the women decided to let common sense rule. Any style was OK until someone's hair actually got caught in the equipment. After that it wasn't a fashion issue anymore but a safety one. Long hair would have to be tied up or cut off. None of the men had long hair, but, Seddon remembers, "there were several guys that came in that had beards. Nobody said you had to shave it. But you had to be able to seal your mask."

Every astronaut is allowed a personal hygiene kit. The men bring razors, unless they are cultivating beards. The women decided they would simply see how things worked out. Women had sanitary needs apart from the men, and menstruation was handled by the individual. Some women opted for hormone pills, to stop their period while in orbit; others did not. Fisher, a mother of daughters, realizes in retrospect that it was odd. "That was never discussed. Each woman dealt with that individually." They never discussed their periods with each other or with the supply crew. The first woman to fly was provided with an excessive number of tampons, and a pair of scissors. No one had the temerity to ask how many a woman might need but someone had stopped to think that once out of the box, they might float off, and so they were linked together like sausages, and needed scissors to separate them.

Some of the women wanted to wear makeup in orbit, others never used it but agreed that those who wished should by all means take it along, as well as a mirror to let them see what they were doing. The only cosmetic all astronauts agreed on, including the men, was moisturizer. The shuttle environment was dry and astronauts applied balm to their lips and wherever else they could reach.

Clothes were another matter. The pilot astronauts were used to getting their underwear from military bins labeled "small," "medium" and "large." But women do not select bras or underpants that way. Women's underwear is very size specific and this presented a problem that was solved by a field trip to a Houston department store. Each woman selected what she needed when she would be on the shuttle, and the order was passed on to NASA. Only a small amount of space was allotted to personal items on the shuttle, and all astronaut clothing belonged to NASA. Just before launch, the women were given the underwear they had selected, with their names on it, and after each flight that underwear was washed and kept by NASA for the next trip. But, Seddon explains, "That was an accommodation. If they had said, 'That's not fair. You need to do small, medium or large, panties and bras,' we'd have done that."

Space suits, on the other hand, were a real problem. As Seddon recalls, "When we came, they simply thought they could take big space suits and proportionately cut them down. . . . The smaller women could never get one of those torsos to fit." Anna Fisher got the job of helping design a suit for women. This had nothing to do with fashion and everything to do with safety. A poor fit is unacceptable because it would make it impossible to move. As a small person Fisher found the unisex stock sizes simply unworkable. "Women are not smaller men," she reminded the astronaut office, which apparently had to be reminded. "Women are built differently." For a while NASA thought it would be more economical to select a large candidate who would fit the suit, rather than adjust the suit to fit a smaller woman, but tall women are also built differently, another lesson NASA had to learn.

All of the astronauts in the sixties had their heartbeat and pulse monitored as they were launched, and sometimes throughout their missions. As far back as the Gemini flights, NASA physicians discovered that accelerated heartbeat and pulse were ameliorated by letting the astronauts get out of their space suits. In the intervening years, NASA had learned that the shuttle crew needed helmets during launch for noise protection and to let them breathe air. They also needed pressure suits for launch and re-entry. With that in mind, shuttle astronauts had to decide what to wear while at work in orbit. The women could choose between a blue jumpsuit with lots of pockets or shorts and a tee shirt.

Television kept the public in tune with the space program. After *Apollo 11*'s live-from-the-moon images, television cameras accompanied every astronaut, almost everywhere. The privilege of going into orbit came with relinquishing some privacy. Each astronaut was taught to use a variety of cameras, and the television cameras were there, and on, during most of their waking hours. Before the first shuttle flew, the press reported the installation of a striped "privacy curtain" separating the toilet from the rest of the workspace.

The toilet, however, was a luxury saved for actually being in orbit. While waiting to launch and strapped into position for hours at a time, NASA provided another option. "One thing we did discuss," Fisher recalls, "was the development of what was going to contain urine for the launch." And since women couldn't use the urine containers that the men used, "they came up with these diaper things. In fact, do you know about all these new absorbent diapers that they have for babies? That's where they came from, because there was no woman that was going to wear a diaper that was like an old diaper. So one of the chemists in our crew systems division discovered that absorbent stuff and decided that's what we would use for females. Then they said, 'Well, why shouldn't we just use that for both sexes?' So somewhere along the way that became standard for everyone to use. So anyway, we did discuss that." Kathy Sullivan remembers their talks and suspects that "if you polled all six of us, I suspect you would find unanimously, there was a confident presumption that biology is pretty resilient. There was a whole similar list of fears in early space days. Would they be able to swallow? Would they be able to urinate?" They were about to find out.

Finally, in April 1981, the first shuttle, *Columbia,* went into orbit. A crew of two tested its design and performance. Four more test flights followed and the new mission specialists knew they would soon fly. As they awaited crew assignments Marvin Resnik believed that the choice of "first" female had narrowed to the two youngest women, Sally Ride and his daughter Judy, both of whom had excellent hand–eye coordination by virtue of either champion tennis play or competitive piano performance. Crew selection, like selection into the corps itself, was largely decided by George Abbey in consultation with the head of the Astronaut Office, who in 1983 was John Young, and the flight commander, Robert Crippen.

Following established procedure, on April 19, 1982, Chief Astronaut John Young announced the names of his crew for STS-7; one of the five crew members on the *Challenger* would be Mission Specialist Sally Ride.

The wait was over. The first American woman to fly in space would be the thirty-two-year-old astrophysicist from California. Ride met the enormous media interest with grace. Although not eager to be the center of attention, she responded to questions with a serious smile. During the six-day mission the crew would deploy satellites for Canada and Indonesia and she would operate the new Canadian-built robot arm system. Crippen explained that Ride had been chosen because of her experience with the robot arm and because she "could get everything she knows together and bring it to bear where you need it." STS-7 left Cape Canaveral on June 18, 1983.

NASA held a party to toast the shuttle as it was about to leave Florida carrying the first American woman astronaut. NASA had asked the Reagan White House to send a list of Republican women to supplement its own guest list, but few Republicans accepted the invitation. Brian Duff, in NASA's public relations office, invited Gloria Steinem and Jane Fonda. Dressed down for inviting the one-time peace activist, he defended his decision by describing Fonda as a good role model for young girls. Perhaps she was. But not in Reagan's eyes. Almost fifteen years had passed since Fonda's visit to Vietnam, but she was still a red flag for Republican politicians. Her presence infuriated the Reagans. Within a short time Duff was "transferred" to a dead-end job, and soon left NASA.

Fonda responded to the brouhaha by commenting on the diversity of the guest list, which included Edward Teller, the nuclear physicist who was the model for the mad scientist in the anti-war film *Dr. Strangelove.* As for the president, Fonda said: "Just because the White House has a 'gender gap' is no reason for NASA to have one, too."

The uproar did not rain on Sally Ride's parade, although the Soviets almost did. Seven months before her flight they sent a second female cosmonaut to *Salyut 7,* their newest space station. The world behind the Iron Curtain celebrated the flight, but hardly anyone in Houston cared. They were caught up in the shuttle program and ignored the "first" records the Soviets seemed determined to set.

A few weeks later, in August 1984, Judy Resnik went up on STS-41D. Then in October Ride flew a second time in a crew with Kathy Sullivan; it was the first mission with two women crew members and the first in which one woman did a spacewalk and looked through the shuttle's windows from the outside. During their flight an Australian journalist had confused Sullivan and Ride and had asked Ride how she felt about being the first American woman to walk in space. Sullivan identified herself as "the other Sally" and, passing the mike to Ride, let her explain that "Kathy" would answer questions about the walk. At Mission Control in Houston a female engineer observed, "They all look alike, right?"

In November 1984 it was Anna Fisher's turn. Then in April 1985 Seddon flew on STS-51D, where a glitch occurred when a satellite went awry and an unscheduled spacewalk was hastily arranged to retrieve it. Inside the cargo bay Seddon fixed the torn sail on the instrument and launched the repaired instrument again. Commenting on the procedure at Mission Control, a spokesman described Seddon's work as "the skill of a good housewife." Ride, who was working the mission as the astronaut assigned as capcom with ground control, corrected him. She pointed out that Seddon's stitches were the work of a heart surgeon. The women astronauts may not have been close friends, but they stood up for each other.

In June 1985, Lucid flew, the last of the astronauts of the class of 1978, male or female, to get an assignment. In November and December, Dunbar and Cleave finally got their chance. By January 1986 all eight of America's women astronauts had flown at least once.

Meanwhile, in Star City, as the cosmonauts congratulated their American rivals, *Pravda* mentioned the possibility of an all-female Soviet crew. That would leapfrog the Soviets program ahead of the Americans in the propaganda battle over which program best understood the value of sending women into space.

5

BACK IN STAR CITY . . .

THROUGHOUT THE SEVENTIES, as American girls fascinated by spaceflight turned to fantasy for models, Soviet girls had a real cosmonaut to revere. Although no Soviet woman had flown since Valentina Tereshkova in 1963, she remained a public figure and spoke eloquently about her three days in orbit, again and again reminding audiences in Eastern Europe and the Third World that she, a Soviet woman, had seen Earth from space.

The Soviet people did not know how harshly Tereshkova's performance had been judged in Star City, and they did not know, and probably would not have cared, that the West dismissed her mission as a "stunt." She had demonstrated to a generation of Russians that girls, too, could aspire to careers in space. By the eighties, more women than ever looked to space for a career. Soviet society had changed, and so had its space program.

A decade had passed since the Soviets ceded the Moon to NASA. Instead, since April 1971, when the Soviets launched their first *Salyut*, they had concentrated their efforts on launching reusable space stations, and setting records for endurance in orbit. They had devised a plan (for the most part in use in the twenty-first century) in which they paired their first station, *Salyut 1,* with a small *Soyuz* "taxi," that delivered the crew to the orbiting station, then docked and remained there. The plan started well, with a three-man team living on the station for three weeks in June. However, believing they were unnecessary because the *Soyuz* was secure, the cosmonauts did not have the usual space suits that included an emergency oxygen supply. When an emergency did occur on their return to Earth, triumph turned into tragedy when the spacecraft depressurized, killing the entire crew.

Chastened, the Soviets spent the next two and a half years redesigning the *Soyuz,* and during these years NASA dominated space with *Skylab.* After *Sky-*

lab was left empty in 1974, the Soviets launched six more *Salyut* stations and retook the lead. The stations did not all remain in orbit for very long; two were military space stations with short, secret assignments. Then, in 1977, the Soviets launched *Salyut 6*, a space laboratory, and in keeping with Cold War competition, established a record when, in 1980, Valerii Ryumin, and his Commander, Leonid Popov, spent 185 days in orbit. Happy as the Soviets were to please their own citizens, the audience they looked to was the rest of the world, particularly the United States.

In keeping with this popularity contest, and with a roomy space station in orbit, the Soviets began launching "guest cosmonauts," visitors who usually stayed in orbit for a week. These included the Cuban Arnaldo Tamayo Mendez, the first black man in space, as well as men from Poland, Germany, Bulgaria, Hungary, Czechoslovakia and, from outside of the Communist bloc, a Frenchman.

In 1982, *Salyut 6* gave way to *Salyut 7*. The last of the series, *Salyut 7* was larger than its predecessors. It could sleep six people at a time if you included those visiting in their *Soyuz* ferry vehicles or *Progress* supply spacecraft and so expand the size of the station. *Salyut 7* could also entertain visitors better than its predecessors because it had retractable shields over the portholes to improve the view—the major free time diversion of most space travelers—and a refrigerator with room for more food since some cosmonauts would be aboard for a long time.

It was at this time that Valentin Glushko, head of NPO Energia, which had taken over Korolev's design bureau under a variety of names until it became NPO Energia in 1974, took control. Glushko had apparently been interested in flying women cosmonauts for some time. In the generation that came of age in the seventies there was a large pool of highly skilled women. Soviet women had continued to study science and engineering, but while some women in this new generation won high-level jobs, they no longer outnumbered their male peers as women had in the years following World War II. As a result of the return to a normal ratio between the sexes, technical organizations, like the space program, did not need skilled women as they had in the immediate post-war years. Equality between the sexes had come to mean that women worked full days, like men, but then, unlike their husbands, they shopped, cleaned and took care of a child with no help. Yet women clung to past promises. Most Soviet women worked, and engineers and pilots especially remembered Tereshkova's flight, and hoped to follow her into orbit.

The ill will toward Tereshkova in the cosmonaut corps had not dissipated by 1978 when Georgii Beregovoi, a former cosmonaut and director at Star City, declared that the social differences between men and women were too great ever to risk launching a mixed crew, or a crew without men. "Women are more emotional and are upset more easily," he stated. Chief Cosmonaut Lieutenant General Vladimir Shatalov, the Soviet air force official in charge of cosmonaut training, described the omission of women in the cosmonaut corps as a chivalrous gesture. He explained that he did not want to subject women to the physical strains of long flights that could "shrink heart size and sap muscle strength."

These men must have exchanged ironic shrugs when, just a year later, Valentin Glushko, who had become not only the new chief designer, but an honorary member of the powerful Central Committee of the Communist Party, reversed policy. Exactly why is unclear except that he, too, was a long-time space visionary and wanted his own list of space "firsts" and to upstage NASA, which had just accepted women into the astronaut program, by launching not simply a second Soviet woman but an all-female crew to the *Salyut.* He initiated a complicated selection process by sending out a notice that he was now interviewing women for a new class of cosmonauts. According to people who had worked with him, Glushko was determined to orbit this all-female crew.

The candidates were selected from different specialties; five physicians from the Institute of Medical and Biological Problems in Moscow; three engineers, one from Leningrad and two from NPO Energia in Korolev; a scientist from the Academy of Sciences; and a test pilot with a degree in engineering. Only five of the women actually trained as cosmonauts and only three were assigned missions.

Irina Pronina, an engineer, was one of the first to respond to the notice. She recalls: "I was in elementary school when Valentina Tereshkova flew and like all little girls I wanted to be a cosmonaut because I understood that women could fly." An only child born in 1953, Pronina grew up in a family that was intimately connected with the space program. Both of her parents had worked in the fifties with Korolev in what Pronina calls "the legendary firm." Pronina grew up near the city of Korolev (formerly Podlipki and then renamed Kaliningrad in the 1930s, but renamed again for the Chief Designer after his death in 1965), where spacecraft were designed and built. "The romance of flights to the stars was common at the time and I didn't know then that my parents worked in the field."

Her father knew the politics as well as the risks of the cosmonaut corps and did not encourage her. Nonetheless, she recalls receiving a gift from him when she reached ninth grade in the late sixties, the *Encyclopedia of Cosmonautics.* "This book was too complex for me," she remembers, "I could not figure out how to master it." But she tried, and loved him for giving it to her, and loved the book for all it contained. At one point, she asked her father, "How can I become a cosmonaut?" He told her to study, do well in school, be good at sports, and have a good disposition.

"So every morning I took a cold shower, did exercises, went skiing for long distances with my father, and cultivated endurance." She also "studied and took part in social activities at school." She says that as a teenager she gradually forgot her childish dream and talks with her father and wanted instead to be clever and beautiful and respected by friends and teachers. But she also wanted to study, as her parents had, at the Bauman Institute—originally the Bauman Higher Technical School, the university Korolev had attended, founded in 1830 and transformed by the Soviets into an analogue of the California Institute of Technology. A generation younger than Valentina Ponomareva, Pronina found warmth and encouragement at home. She did follow her parents to the Bauman, where she studied aircraft design. "After that," she smiles, "the book my father had given me was too simple."

In 1979 she had decided that she did, after all, want to be a cosmonaut. She knew, of course, that no woman had flown since Tereshkova, but she had no idea why. She had never heard of the other women who had been commissioned with Tereshkova. She had assumed that Tereshkova had always been alone, and that her flight had been a great success.

When she saw Glushko's notice, she immediately submitted, as requested, a simple letter of intention outlining her qualifications. Then she waited. The selection process was very slow. Pronina believes that 2,500 women from institutions throughout the space industry responded, and these were just her fellow engineers. There was a separate selection for medical cosmonauts. The yearlong process included a series of physical examinations and recommendations from her superior at the lab. There was also, she recalls, an evaluation of her looks and behavior as a Soviet citizen. This last requirement meant letters of approval from the local Communist Party cell. Not until a full year had passed did she learn that she had been chosen, along with two other engineers.

One was Natalya Kuleshova, who, like Pronina, had parents involved in the space program. Kuleshova was born in 1956 in Kubinka, near Moscow, where her father was studying at the Rocket Forces Academy. With his degree he went to work at the Cosmodrome, the launch facility in Baikonur. She longed to tell him about the competition but, like much else in the Soviet Union, everything about the military was secret, and the astronauts were part of the Air Force. Eager to get into the program in any guise, she had applied to be an "engineer tester" rather than a cosmonaut. When she did tell her father—against regulations—that she was, in fact, applying for the cosmonaut team, he responded unsympathetically: "Women have no business there."

In spite of his attitude, she remained a candidate. It was a long process. Kuleshova remembers the "triangle" of signatures she had to get: one from the secretary of the local Party cell, one from the secretary of the local Komsomol, the communist youth organization, and one from the head of the local trade union organization. At some point other candidates were eliminated from consideration because of their ages or ethnicity. What this may have revealed, she suggests, was a search for the most "Russian" candidates, which meant the exclusion of Jewish women and women from the non-Russian Soviet republics. All the finalists were ethnic Russians. The age limit was capped at thirty-three, perhaps, she suggests, to remove Tereshkova and the other women in that class (whom she learned about later) from trying to slip back in.

Svetlana Savitskaia, another candidate who was a test pilot with a degree in aviation, is more certain that was the case. She had been Glushko's student and knew him well; he had told her that almost all the women in the first group from 1962 had asked to be considered for this second women's flight. But he had observed that while they were gold, they were too old. When she mentioned Tereshkova, she wrote, he gave her a penetrating look, paused, and said, "Tereshkova is a national treasure. Space flight is dangerous." To protect her, someone else would have to take the risk.

Irina Pronina and Natalya Kuleshova received a mixed welcome when they joined Savitskaia in Star City in 1980, just as Tereshkova had in the early sixties. Star City had not changed very much. Svetlana Savitskaia, who was part of the group, says that twice during the training period the women were told to leave. Both times the previous evening Beregovoi, the director of Star City, had "dropped in" ostensibly for tea, and the next morning they were notified that the program had not been finalized and must now be stopped. They were told that a bus would arrive soon and they should pack up and leave. Savitskaia writes that they then informed Glushko, who told them to remain at the office of Energia. Glushko was telling them to "stand by" until they were allowed to continue training at Star City.

Pronina recalls that until the day she arrived in Star City she had not felt that being female had had any impact on her career. Looking back, however, she realizes that in the Soviet space program gender discrimination began "at the stage of crew formation." Was it a shock to her? No, she recalls: "Unpleasant, but I always believed in a better future."

The Iron Curtain, she explains, had not shut out news of political changes in the West and Western ideas had sifted into the Soviet Union. The women's movement that began in the United States in the sixties had spread like a backward dawn from the west to the east, bringing light and waking people up. Pronina says that she knew that women were demanding equal access to jobs and universities in the West and she believed that the movement would eventually have an effect in her country, but not in time, she says resignedly, to help her fly in space.

Pronina and Kuleshova were soon accepted as candidates for the new group of cosmonauts, each of whom, Pronina believes, Glushko personally interviewed. The finalists included three engineers, one scientist, four doctors and one pilot—who was also an engineer. Kuleshova concurs that Glushko himself made the selection for a planned all-female crew, perhaps to showcase Soviet confidence in women.

But Glushko's plans accelerated in 1982 when NASA announced that Sally Ride would fly on STS–7 early in 1983. Suddenly he had a deadline. If the Soviets were to preempt NASA, they had to send a woman up quickly. This meant that plans for the all-female flight were postponed—as it turned out—permanently. The change meant that the chosen woman cosmonaut would have to fly with two men because cosmonaut center rules demanded a flown cosmonaut, not necessarily the commander, on every flight.

The all-female crew would have been a public relations bonanza. But it would have interrupted the launch schedule that the cosmonaut leadership had carefully planned and was under pressure to complete. The women were not part of those plans, and the male cosmonauts were not interested in public relations, at least not when it put women ahead of themselves.

At this point, however, in a departure from form, the Soviets announced a new crew for *Salyut 7* while the old crew was still on the station. TASS told the world that a group of women cosmonauts was in training, of whom two were being prepared to fly—a pilot-engineer and a flight-engineer. But with Sally Ride scheduled to fly, Glushko had to correct the misfired announcement of an all-female crew and propose a new plan that would please his friends in the Kremlin and at Star City. Instead of the all-female crew, the Soviets announced that they would send a second Soviet woman into space before there was a first American.

Salvaging what he could from his original plan to get Soviet women aloft before Sally Ride, Glushko declared that he would send one woman in a mixed crew. Like Ride, she would be the only woman in a crew of men. It was not the dramatic gesture he had wanted to make, but being a realist, he would do the possible right away.

From the candidates, he selected Savitskaia, the aviator with a graduate degree in engineering. He announced that Pronina and Kuleshova would be her backups. Like the women in Tereshkova's group in 1963, Pronina and Kuleshova were promised another opportunity in the near future and they were assured that this hastily arranged flight was simply added on, that their all-female crew, as originally planned, would be rescheduled.

As for the choice of Savitskaia, unlike the selection of Tereshkova, which was suggested by a committee but was ultimately resolved by Khrushchev, Glushko made the decision alone. Neither of her backups believes that the decision could have gone any other way. If there had been a women's crew, they would have flown with her. But alone, she was far and away the front-runner. This meant that Savitskaia would be the only woman eligible to command a crew.

Savitskaia was in every way the obvious choice. She was famous abroad as well as in the Soviet Union. The British press had embraced her in 1970 and dubbed her "Miss Sensation" when she set a record at the Hullavingston Air Show and won the World Championship in aerobatics as a member of the Soviet National Aerobatic Team. She was the only daughter of Air Force marshal Yevgenii Savistsky, two times a hero of the Soviet Union for wartime bravery and part of an elite with roots in the Stalinist thirties.

But apart from this pedigree, she was an excellent pilot and represented the highest possible level of aerobatic skill. She had a superb education. Raised bilingually in English and Russian, she excelled in swimming and figure skating and as an aviator she had begun setting records when she was seventeen by skydiving from 14,252 meters. She came from a more connected family than the women she was competing with, and was even more distant than they were from the pre-Revolutionary bourgeois background that may have handicapped Valentina Ponomareva twenty years earlier.

Although the United States had opened its military academies to women in 1976, two years after opening its flight schools, no American woman aviators had progressed far enough by the early eighties to apply as a pilot astronaut. Unlike her American counterparts, Svetlana Savitskaia—a graduate of the Moscow Aviation Institute—was a test pilot who had flown with the Soviet Air Force.

The flight plan called for a research cosmonaut, and although Savitskaia was not a scientist, she had studied engineering and was prepared to perform scientific tasks during the eight-day mission. Most of the science she would be doing entailed monitoring her own physiological responses to weightlessness. Most important to the Kremlin, she was already known at home and abroad. She was pretty enough, with an attractive smile when she deigned to show it, and she had already proved that she was tough.

She would have to be. It is hard to imagine what life was like for her during that week on *Salyut 7.* Before the launch she had told a press conference that the future of all women in the space program depended on the quality of her performance. This was a veiled dig at Tereshkova, for by this time Savitskaia had spent months training in Star City and had heard all the negative rumors about that earlier flight and knew that many cosmonauts considered Tereshkova an embarrassment. Tereshkova, for her part, was remarkably unsupportive of Savitskaia. She never showed up during the training period or at the launch in Baikonur. Some news stories suggested that Tereshkova was having health problems and could not attend; others suggested that it was simply sour grapes. She had wanted to fly again herself.

Without Tereshkova's blessing, and convinced that the future for Soviet women in the space program depended on her, Svetlana Savitskaia took her place in the left seat in the small *Soyuz* spacecraft on August 10, 1982. She wore a light space suit identical to those worn by her neighbors: on one side the commander, veteran cosmonaut Leonid Popov; on the other another rookie like herself, Aleksandr Serebrov.

Psychologists at Moscow's Institute for Biological Problems are generally proud of the psychological studies they do to ensure harmony between cosmonauts in space. In this instance they surely failed. They did not seem to have paid attention to Serebrov's outspoken hostility toward women cosmonauts. The Soviet media seemed equally insensitive. In a televised interview before the flight, Serebrov asked, somewhat ingenuously: "Why would a woman want to do a man's job?"

Greeting the trio when they docked with the *Salyut 7* were two very experienced cosmonauts—Anatoly Berezovoi and Valentin Lebedev. The *Salyut* was roomy compared with the cramped *Soyuz* and with the *Soyuz* linked to it as an extra compartment, it was luxurious by Soviet standards. When Savitskaia floated into the station from the *Soyuz* to begin her weeklong visit, *Salyut 7* had been Berezovoi and Lebedev's home for a month. This was enough time for them to grow a bouquet of arabidopsis (flowers from a scientific experiment), which they presented, as is the Russian custom, to their female guest.

Their next gift was also traditional. "We've got an apron ready for you, Sveta [a diminutive of her name]. It's as if you've come home. Of course we have a kitchen for you. That'll be where you work." The Soviets described the greeting as a joke, which it was, but a joke with roots in a culture where a woman's role was still defined by housework, in addition to everything else.

This gesture was referred to by a Leningrad psychologist, V. Garbuzuv, who had written in a trade union newspaper, *TRUD,* that "Giving a girl an apron as a present before she is given a Sunday dress instills in her the skill of housekeeping, the skill and taste for this eternal female cause."

Savitskaia was ten years older than Tereshkova had been during her flight, and because of her aviation records and accomplishments she was self-confident beyond her years. Besides, her father was still deputy commander of the Soviet Air Force—which made it unlikely that anyone in the space program dared disparage her to her face.

But there were other ways to get at her once she was on the space station. Serebrov was antagonistic throughout the flight. He disliked her, and the veteran cosmonauts, her hosts on the station, were unhappy, too. They resented having to entertain any visitors, and they especially resented her. We know this because with the end of communism in 1991, two written accounts of her visit became public. The first was Lebedev's diary. He describes the initial meeting of the two crews, noting that as soon as the *Soyuz* had docked and the hatch between them opened, the two men floated in right away—as was expected—but Savitskaia was busy indulging in a singular act of feminine vanity—she was combing her hair. (This is, of course, a gesture that would have been wasted in microgravity.) Later that day, at dinner, her new blue apron in hand, her hosts made it clear that being a cosmonaut and a test pilot did not free her from the female role of preparing dinner. She agreed, probably because she was good-natured and because there wasn't really much else to do.

Berezovoi was equally resentful. He confided in a letter to his wife that was auctioned at Sotheby's in New York in 1993 that Savitskaia's presence disturbed him: "Well, isn't there a superstition among sailors that a woman aboard a ship is a bad omen? At the moment I don't know for sure, but I can't imagine all the difficulties that we will have with this mixed crew. . . . I'm afraid this will not be limited to just the problem of shared facilities. . . . I will say nothing of Svetlana. She is a woman and that says it all. It will not be easy."

This was the private face of their feelings and actions. To the world the crew reported that everything was progressing ideally.

In fact, as Berezovoi wrote in the rest of his letter, Savitskaia and Serebrov were sparring, and she was following orders to the letter, but no more. "Serebrov and Savitskaia are like a cat and a dog. Sasha [diminutive of Alexander] has turned out to be one of those people who like to whisper about their friends, in this case about Svetlana. Savitskaia isn't exactly going out of her way to do anything over and above her program description. Today a lot of mistakes were made. In broadcast reports we stay ever cheerful and lie a lot. In general, the presence of a woman greatly limits the freedom of the station and complicates daily life. We have assigned her one of the two transport ships for physical needs, but she sleeps with us inside the station."

When she returned, Savitskaia attracted crowds of reporters as Tereshkova had done twenty years earlier. But unlike her predecessor, Savitskaia did not seem to have made an effort to be charming. Like the American women who would follow her within a few months, she deflected all questions regarding how being female had made a difference in being in space or doing a cosmonaut's work. She presented herself as just one of the crew.

Her cool demeanor did not fool the international press. They homed in on the flowers and the apron as a way of exposing flaws in Soviet culture. Instead of bolstering the image of the Soviet Union as a haven for ambitious women, the rushed second flight backfired. By 1982 most of the Western press was sensitive to women's political and social ambitions. They were also sensitive to the slights that women experienced as they struggled for equality. These men and women, their consciousness raised, were unwilling to let the Soviets get away with grandstanding one woman, and that with mixed messages, while making life hell for so many others.

The New York Times, in particular, took *Izvestia,* the communist daily newspaper, to task on its report on Savitskaia's week in space. *Izvestia* had described her as "charming and soft, a hospitable hostess [who] likes to make patterns and sew her own clothes when she has time to spare." The Soviets commented on domestic behavior that emphasized the differences between men and women—which had nothing at all to do with her performance as a cosmonaut. Georgii Timofeyevich Beregovoi, director of the space center, suggested that "the presence of a woman exerts an ennobling effect on the microclimate of the small group." He noted that Soviet "Specialists had pointed to the high efficiency of all three cosmonauts, and we believe it is to a certain extent a result of the fact that a woman is working a crew together with men."*

The American press leaped on the overt Soviet hypocrisy. True, women made up more than half the Soviet labor force and comprised 70 percent of all doctors and teachers. They also filled high, if not top, positions in engi-

* It's perhaps worth nothing that Beregovoi did not mention the toilet, and neither did anyone else, in contrast to the Americans, for whom toilet design was always an issue. The Soviets had a unisex toilet that worked well and was never, apparently, a problem. Feminist demands for women in space notwithstanding, Americans inside and outside NASA were still debating toilet design, even as the Soviet station with its single toilet orbited the planet.

neering, and photographs of the teams responsible for Soviet spacecraft include many women from the beginning in 1957. But women also dominated low-level jobs like street sweeping. Women filled positions throughout the Soviet workforce, with many more at the bottom.

There was also the matter of tokenism. The only woman who then headed any Soviet organization, the Soviet Women's Committee, was Valentina Tereshkova. There were no women in the Politburo, the Soviet's top government ministry. Soviet women were supposed to receive flowers and chocolate, but not the privilege of driving a car.

At a press conference immediately upon her return in 1982, Savitskaia tried to counter the condescending attitudes about women that were common among the male cosmonauts. She pointed to the results of the medical studies she had done in orbit that indicated absolutely no differences between the way she, a woman, responded to space compared to men. They had tested her heart and lungs before and after the flight and found her fit. The Soviets knew as much about her, a single data set, as they knew about their male cosmonauts who had spent equivalent time in orbit. She seemed fine, and that's what counted.

As for her performance of specific tasks, Savitskaia felt that she had performed better than her male colleagues and suggested that more women should be sent into space. Although she never specifically mentioned the women she had trained with, she was clearly referring to them. In the summer of 1983, the Soviet Space Agency still planned to fly Pronina. That September, a fire destroyed a *Soyuz* only seconds before it would have been launched to rendezvous with *Salyut 7*. The cosmonauts escaped, but within minutes the aborted *Soyuz* had set fire to the Baikonur Cosmodrome. In that inferno, the entire launch schedule was shattered.

Before a flight could be arranged for Pronina, NASA announced that Sally Ride would be making a second trip in 1984, taking along fellow astronaut Kathryn Sullivan, who would be the first woman to perform an Extra Vehicular Activity, an EVA or spacewalk. The Soviets responded by scrubbing Pronina's flight in order to send Savitskaia up again—this time to do an EVA. The trip would be a double strike. The Soviets would preempt Ride's chance to be the first woman to return to space, and Sullivan's chance to be the first woman to leave a spacecraft to take a walk.

On July 18, 1984, Svetlana Savitskaia left a new launch pad at Baikonur for a second visit to *Salyut 7*. Her second flight differed in almost every way from the flight two years earlier. Due to a change in protocol, the Soviet Space Agency now officially insisted that each new cosmonaut have experience as a flight engineer before becoming a commander—something like NASA's requirement that pilot astronauts on the shuttle fly twice as pilots before assuming controls of the craft as commander. In 1984, the Soviet space team was eager to use Igor Volk, an experienced test pilot, on a later mission, but he needed space experience. They had also committed themselves to flying Savitskaia a second time. They solved these two problems by adding an extra flight to the tight schedule, bringing Volk and Savitskaia to the *Salyut* together. This time Savitskaia, who had already flown, sat on the right as flight engineer and Volk rode in the center seat as the rookie research cosmonaut. Happily for Savitskaia, they had been in test pilot training together and were old friends.

The commander of the crew they joined on the *Salyut,* veteran cosmonaut Vladimir Dzhanibekov, did not, as far as is known, complain to anyone about the added visitors, at least not in writing. Two days after leaving Baikonur, on July 20, 1984, Savitskaia floated through the hatch of the station into space as Dzhanibekov, her partner on the spacewalk, moved aside to let her go first. Thus began the first EVA by a woman. Once outside, the two lost no time in getting on with the job, which was to test new tools. First Savitskaia cut, welded, soldered and took pictures, then Dzhanibekov did the same. Three and a half hours later, they returned to the space station congratulating each other on their skills and success. TASS, the Soviet press agency, reported that Savitskaia's performance "has vividly shown a possibility of women's effective activity in the performance of complex research work, not only aboard manned orbital complexes but in open space as well."

Savitskaia returned to Earth on the 29th of July and held a press conference in lawn chairs in the grassy pasture where she had landed, still in her space suit. She emphasized the beauty of the Earth and the importance of respecting it. But when asked what the most wonderful moment of her trip had been, she did not mention the stars or the experience of being outside the station in open space. Rather she marveled at the wonder of the plowed soil when she landed on the steppe. "The hatch was opened and all the smells rushed in. That moment was really nice."

What was not so nice, she made clear in the next few days, were some references to her presence as a kind of tonic to cheer up the men she was working with. "We do not go into space to improve the mood of the crew," she said. "Women go into space because they measure up to the job. They can do it. . . . At a minimum," she warmed to the subject, "women are equal to men in space. Women are actually better at some space tasks than men. They are better at dealing with precision tasks. They are more meticulous. They are more flexible at switching from one task to another. Men, of course, are better where heavy exertion is required." Savitskaia saw women as better, not worse, than men, and certainly different.

Whether or not Svetlana Savitskaia would be leading many more Soviet women into orbit, she was feted for what she had already accomplished. She received a second Order of Lenin and Gold Star to add to the one she had received for her first flight and eventually, after the end of the Soviet Union, won a seat in the Russian assembly, the Duma.

In 1984 Pronina and Kuleshova returned to waiting. They were still promised an all-woman crew, which they assumed Savitskaia would command. Glushko still talked about such a flight because, among other reasons, it was something that NASA had not considered. This could be another "first" for the Soviets to parade before the world. Building on Tereshkova's historic flight, they could remind the world of communist egalitarian treatment of women by sending an entire female crew to a space station sometime in the mid-eighties.

Glushko selected a crew, but power problems on the space station caused delays. Then Savitskaia took a leave to have a baby, which left Star City without a woman who had already flown and would qualify, under the new rules, to command the flight. The women remained at Star City, training to fly on the new Soviet space station *Mir.* On the other side of the world, NASA kept the shuttle flying with one and then two women in a crew. The prospect of an all-female flight began to seem old-fashioned. Finally, after the fall of the Soviet Union in 1991, the Russian space agency retired Pronina and Kuleshova, ending their hopes of spaceflight.

Pronina and Kuleshova are not as philosophical about their experiences as were the women in Tereshkova's class. Most of the men they trained with eventually made it into orbit. They still resent the hypocrisy of the Soviet government, which had tricked them into believing the propaganda about equality of opportunity. In the early eighties the Soviet Union looked as if it would last forever. Women like Pronina and Kuleshova, who benefited from its privileges, have had a hard time coming to grips with the way the system had helped them with one hand and taken away opportunity with the other.

6

TWO AMERICAN WOMEN

THE NEW YEAR OF 1986 blew in cold at the Kennedy Space Center. To January's visitors, it might as well have been Moscow as they waited, bundled in down jackets, to watch the first shuttle launch of the year. STS-61C, *Columbia,* carried an all-male crew that included Florida congressman Bill Nelson as a payload specialist. After seven delays it finally got into orbit on January 12 and the crew deployed an RCA communications satellite. But the cumulative delays had pushed back the departure of the next in line, the shuttle *Challenger,* to January 24.

Each shuttle's mission is special in its own way, and although no one knew it at the time, *Challenger* would change NASA forever. By the last week in January 1986, two Soviet women and eight American women had flown in space. Women astronauts were no longer a novelty. But *Challenger* carried a female payload specialist who was not a politician or a visitor from another country but rather a "citizen" astronaut who would make *Challenger,* STS-51L, a milestone for ordinary Americans. She was Christa McAuliffe, a schoolteacher who, once selected, became an instant celebrity. Zipped into her blue jumpsuit, she was living proof that a woman who taught children in school could fly into orbit. Her very public flight—and the media circus leading up to it—would put NASA in a spotlight that spilled over into every American classroom.

Now, with temperatures even colder than earlier in the month, it was *Challenger's* turn to wait. Its launch was delayed first by bad weather in Dakar, Senegal, where the flight was supposed to abort if it ran into trouble. Then it was the weather in Florida, and after that seemed to have improved there was a problem closing the hatch door. High crosswinds at the Cape called for yet another delay. While NASA waited, fern and strawberry growers who rented government land near the Space Center rushed to turn on sprinklers so that a film of ice would insulate their plants against the forecast freeze. It wasn't the coldest winter on record in central Florida, but it was colder by 15 degrees than it had ever been for a shuttle launch.

NASA wanted to get this mission into orbit. Nineteen eighty-six was the year the agency expected to begin sending the shuttle up frequently and on a schedule like the almost-commuter vehicle it had promised Congress in 1972. Since the first launch in 1982 there had been only twenty-four flights. A decade earlier NASA had projected forty-nine missions a year, a number that, if successful, would have drastically lowered the estimated price for a payload, whether it was a communications satellite or a chemistry experiment, by a factor of ten. There was no explicit pressure from the White House, but the press noted the promised schedule and sometimes joked about the delays. There was a general feeling at NASA that the "show must go on." This flight was especially important for another reason. After three full years of shuttle missions, NASA could finally boast a crew that included African-American and Asian-American men along with two women, a professional astronaut and a schoolteacher. It was the very picture of America.

Like the crew that had directly preceded them, *Challenger*'s crew endured three delays in five days. The cheering squad of friends and families thinned out as the wait dragged on. Still waiting and watching was the crew of the next shuttle mission, STS-61D, *Columbia,* which would be carrying Spacelab, a joint effort between NASA and the European Space Agency. (It would be flown as STS-40 in 1991.) Spacelab was a self-contained module that fit into the cargo hold and would make STS-40 the first shuttle mission dedicated entirely to the life sciences. It had been scheduled to fly ahead of STS-51L (hence its lower number), but because of problems with the organization of laboratory materials, the flights had flipped places in the lineup. But the crew of STS-40 still had a proprietary feeling for that late January slot.

Challenger had different scientific, as well as commercial, goals. It was prepared to deploy Data Relay Satellite-2 designed by TRW, as well as astronomical devices that would observe the tail and coma of the 1986 visit of Halley's Comet. One of the payload specialists, Greg Jarvis, an engineer from Hughes, would study fluid motion in orbit. The other payload specialist, Christa McAuliffe, would deliver school lessons, and the other woman in the seven-member crew, Judith Resnik, a mission specialist on her second flight, would help with all of these tasks.

The two women had been born within a year of each other at the end of the 1940s in middle-sized American cities. They were both adored firstborn daughters, devoted to younger siblings, raised in religious homes and personally determined to succeed at whatever they tried. In the sixties, when early marriages were applauded, both had married classmates. But that's where the similarities ended. When the professional astronaut met the high school teacher, there was little reason to suppose they would become friends.

Judy Resnik, an electrical engineer, was one of the first women selected as an astronaut, yet she never became a household name. Even today, when the word *Challenger* has echoes of the word *Titanic,* few people recognize the names of its crew members—except for Christa, "the schoolteacher," and are startled to learn that two women perished in the tragedy.

Judy Resnik and Sally Ride were the youngest women in the astronaut class of 1978 and Resnik's father believes that, for a while, it was a toss-up as to who would go first. Who could tell how they would choose? Judy had told her Dad that nobody knew. "Most likely they flipped a coin," she'd quipped. "So with a flip of a coin it would have been Sally Ride on *Challenger* instead of Judy." Marvin Resnik had clearly rehearsed this scenario many times. For if it had been a matter of a coin toss and Judy had gone first, she probably would not have been on the roster for STS-51L. But the selection was probably more deliberate than a coin toss, and it is likely that Resnik's intense dislike of publicity precluded her from being the front-runner. Ride certainly never craved the attention she got, but she responded gracefully when it was forced upon her.

Marvin Resnik's retirement home is half a continent away from Akron, Ohio, where he once practiced optometry, sang as a visiting cantor in area synagogues and raised his daughter. He brought a lot of Akron with him to the sprawling house in northern San Diego County. Along with photographs of his son and stepdaughters, the house is filled with memories of Judy. Her piano dominates the living room; her scholastic and professional honors line the walls; and he is happy to describe the woman who kept herself hidden so well from the public.

None of the astronaut women liked the press, but Judy Resnik was particularly uncomfortable with reporters. Loyal and obedient to NASA, she accepted the part of her job that meant talking about NASA to lots of people. The first women were especially popular as speakers and some of them suffered "media shock." Mary Cleave, who joined the astronaut corps in 1980, explained: "All females had to get unlisted phone numbers. There were guys that would call you up in the middle of the night, wake you up to talk about things. . . . You don't understand the value of your personal privacy until you lose it. And it's really shocking. This is part of the hero thing I wasn't prepared for at all." Each woman dealt with it in her own way. Judy Resnik responded by closing up. This was not always easy. NASA could tell the press, as they told the Selection Committee, not to ask women questions they wouldn't routinely ask men. But NASA could not control the press entirely, and they could not protect the astronauts from prurient, condescending, personal or simply stupid remarks.

When asked, Resnik simply told them facts. She grew up in an observant Jewish home in Ohio, where her father had joined his four brothers, sister and parents who had emigrated from Israel after the 1948 War of Independence. In Ohio he had met and married Sarah Polen, a legal secretary, with whom he had a daughter, Judith, and then a son, Chuck. As a child, Judy was surrounded by a large extended family and, with her brother, was the focus of her parents' lives. Her mother organized Judy's days around after-school lessons—piano lessons, cooking lessons, Hebrew lessons and homework. She went to temple regularly and on Friday nights joined the larger family at Sabbath dinners. An able student, she was in the second graduating class at Harvey S. Firestone, a public high school with expensive extras bequeathed by the legendary Firestone tire family. These extras included an indoor swimming pool, well-equipped science laboratories and a planetarium. Few public schools had these luxuries in the sixties, and Firestone immediately won a reputation for being a school for rich kids. The Resniks were not rich, but the children were academically privileged.

Life at home was less comfortable. The Resnik marriage gradually deteriorated, and in high school the adolescent astronaut-to-be rebelled. She let her hair grow long—her mother had kept it in a pert "pixie" cut—and hung out with boys from a less well-endowed school and cruised in a Buick station wagon, regularly swinging by the Skyway drive-in, drinking malts and letting her waist-length hair flow behind her. These boys seemed wilder than her friends in the math club at Firestone, where she was the only girl. But cruising did not stop her from keeping an A average. Nor did her battles with the court following her parents' divorce. Her mother was awarded custody, but Judy took the initiative and asked the court to let her live with her father during her senior year, an arrangement that understandably strained her relationship with her mother. Soon Marvin Resnik remarried; his second wife, Betty Resnik, brought twin daughters, nine years older than Judy, into Judy's life.

There she was in 1966, a "brain" in a generation in which most brainy girls felt that they had to hide their talents. She hid nothing. She was a member of the National Honor Society, the math and French clubs, and had good friends, but none among the high school elite that included the cheerleaders. All the while she continued playing the piano, and when it was time for college she applied to the Julliard School of Music in New York and to Carnegie Tech (now Carnegie Mellon University) in Pittsburgh. Accepted at both, she chose to be one of the few women at Carnegie Tech.

Resnik entered as a math major but soon switched to electrical engineering, one of only three women in the major. She found a friend on the faculty, Angel Jordan, then dean of the engineering department. He recalls her dropping by his office with her dog Elissa on a leash and calling out, "Angel, I am here. Do you have a moment?" She would then settle into long conversations about almost anything that was on her mind. At Carnegie Tech, where she lived in the only women's dormitory, she played the piano for hours in the dormitory living room. She also joined a sorority, was runner-up for homecoming queen, and found a boyfriend—another electrical engineering major, Michael Oldak. Classmates describe them as inseparable.

Throughout college Resnik did not go to classes on Jewish holidays, and as soon as she graduated she married Oldak in an orthodox Jewish ceremony. A decade later, like her father, she had stopped practicing Judaism. He believes that she simply lost interest in religion, but insists that she never stopped being Jewish. Mostly, he says, she hated being categorized. When she declined requests for an interview with *Hadassah,* a Jewish women's magazine, she explained, "I am an astronaut. Not a woman astronaut. Not a Jewish astronaut. An astronaut."

After graduation, Resnik and her husband, both with degrees in engineering, found jobs doing research at RCA in New Jersey. The next year, however, Oldak entered law school in Washington, D.C., and Resnik transferred to RCA's northern Virginia office to be near him. Her interests were changing and she soon left RCA for a different kind of scientific job at the National Institutes of Health in Bethesda, Maryland. She began biophysical research and entered a doctoral program in bioengineering at the University of Maryland. Yet she had not found the kind of work that fit her talents. One day she asked Jeffery Barker, a colleague at NIH, "Where do you get your ideas from?" He recalls telling her that if she had to think about where biological ideas come from, she ought not to be there. "It's the kind of question somebody who is very adept at carrying out someone else's ideas asks. And she was extremely talented at carrying out complex tasks." By this time she had completed her doctoral thesis on the effect of electric currents on the retina of the eyes, something known as dark adaptation. At this point, after six years, her marriage ended.

That was the summer of 1977. She had not, to her father's knowledge, showed any interest in the space program until she was at the National Institutes of Health. "Then they put a sign on the door that said NASA wanted scientists. And they didn't mention men or women. So she called me. She said. 'Daddy, I'm going to apply to NASA to become an astronaut.' I said, 'OK, you'll make it.'"

That summer, Resnik investigated NASA and the astronaut corps with the same thoroughness she applied to everything in her life. She decided it was what she wanted to do, and determined to become an astronaut, she went to Washington where everyone was celebrating the bicentennial. One of the festivities was the opening of the National Air and Space Museum on the Mall. Resnik entered the enormous glass-covered hall where Charles Lindbergh's *Spirit of St. Louis* hangs suspended and examined every display about astronauts, looking for clues about how they had been selected. Then she found the U.S. senator from her home state, John Glenn, who had reversed his views about the merit of women in the astronaut corps. He told Resnik about the job and remained a friend throughout her life.

She left Washington in the fall of 1977 and moved to Redondo Beach, California, to begin work at the Xerox Corporation. She had not forgotten her application to NASA, but it was not something she could count on. However, when she learned she was one of the final 200 women candidates, she began taking flying lessons. Meanwhile she discovered during a routine physical at Xerox that she had a thyroid problem. Undeterred, she went directly to a specialist for a closer examination, and when the condition was declared negligible, she asked for a letter to prove it to NASA. On December 30, 1977, she heard from NASA that she had been accepted to the first class that included women astronauts. She would report to the Johnson Space Center and go on the government payroll, earning $30,000 a year—about what she could earn in industry. She called her father right away and told him she had been accepted, that the news would be announced on January 1. On New Year's Day 1978 Resnik telephoned everyone else who was important in her life, sent in her resignation to Xerox, packed her belongings and left for Houston.

Once settled, she gave her family the impression that she had come home. She liked everything about being an astronaut—the mission of space exploration, the job and her colleagues. She believed that these were people who understood that excellence is a result of hard work and who understood the value of talent or even genius. Astronaut training is expensive and NASA expected its astronauts to remain in the program for at least six years. But Resnik told friends she would stay even longer, as long as she could. She had found the place she wanted to be. Her stepmother believed that until she became an astronaut Judy had always been brighter than everyone else. But in Houston "she had found what she called 'her family.'. . . I think those had to be the happiest days of her life."

As an astronaut she did whatever she was asked to help NASA's mission, which included joining Tom Brokaw in April 1981 in a jerry-built NBC studio at Cape Canaveral. She was there to help narrate the first, and much delayed, launch of the shuttle *Columbia,* to explain to viewers how rockets work and to parry Brokaw's small talk. She met questions like "What do you say when you meet a guy and he says, 'You're too cute to be an astronaut'" with "I just tell them I'm an engineer." Perhaps he didn't ask the right questions. If he had, she could have told him that becoming an astronaut is a lot like becoming a musician. There is an old saw about the lost concertgoer in New York who asks a passerby how to get to Carnegie Hall, and receives the answer: "Practice!" Practice and discipline had been drilled into Resnik as a child; it was the way she lived as an adult. She practiced the piano as she practiced cooking, and friends savored her invitations or her contributions to potluck dinners. At the Space Center she spent hours at the simulator, practicing for her flight.

Those close to Resnik were not surprised that she became an astronaut. They were used to her winning. A few might have been disappointed that she was the second, and not the first, American woman to go into space, but there is no evidence Resnik felt that way. Sally Ride was her friend, and Resnik was pleased for her. Besides, she disliked what she considered the false mystique of being "first." She once explained to an interviewer from the *Washington Post:* "I've been the only woman in my profession and education for years. It is not a new experience. It is important for women to recognize that we cannot stand alone in the limelight as we one by one penetrate areas new to us. True, we must continue forward with our endeavors and our firsts, and broaden our horizons at every opportunity. But firsts are only the means to the end of full equality, not the end itself."

She was the second American woman to orbit the Earth, and that trip had its own unpleasant "first." At T-minus 4 seconds before blast-off, with all three engines humming, Resnik and the rest of the crew heard a loud bang. Fire flashed out from the back of the shuttle, at which point an alarm rang, followed by the high-pitched whining of engines turning off. It was the first launch abort in shuttle history. In a final insult, the fire had triggered the automatic sprinklers, drenching the crew as they climbed out wet and shaken. They had been saved by NASA's computers, which had sensed a malfunctioning valve and ended the launch. The crew had two months to regain their composure and rest, and then they tried again. The abort had been disappointing, but it strengthened Resnik's confidence in the system. The built-in redundancies proved to her that NASA took care of its astronauts.

During her eight years with NASA, Resnik's life revolved around the corps. She was seeing a fellow astronaut, Frank Culbertson, and loved to entertain his family, and they found her fun to be with. There were lots of ways to make her laugh, one of which was referring to actor Tom Selleck, whose life-sized photograph on the back of the "privacy curtain" around the shuttle toilet was a surprise gift from her crewmates on her 1984 mission. She had invited her ex-husband, a good friend, to her first launch, and had carried a sign for her "best friend" that read "Hi Dad," which she made sure the television camera sent back to Earth. Smart, funny, hospitable, on occasion prickly and almost always private, she rationed her smiles in public, bestowing them only when she felt one was warranted. That smile, captured in NASA photos, was worth waiting for.

Like Judy Resnik, Christa McAuliffe had a wonderful smile, but Christa offered hers easily and often. It was one of the qualities that won over the selection board. Sharon Christa McAuliffe (the Sharon was lost from her name before she could talk) was born in Boston, Massachusetts, in 1948, the first of Grace and Ed Corrigan's five children. A veteran of World War II, Ed Corrigan started his second year at Boston College the day after Christa's birth. The Corrigans were on a tight budget as more children arrived, but Corrigan soon got a good job with the Jordan Marsh department stores. Although Grace Corrigan did not return to her brief modeling career after Christa was born, she had liked it enough to send Christa to modeling school. After a few years the Corrigans moved to Framingham, a suburb west of Boston, and there, in her seventh-grade classroom, Christa watched on TV as Alan Shepard rode a rocket into the atmosphere. Like Rhea Seddon, her contemporary, she was fascinated by space travel. But unlike Seddon, McAuliffe never thought she could go there. In fact, she never thought she could do anything adventurous. She wrote in her application to NASA that a career counselor had told her junior high school class, "The only thing a girl could be was a nurse or teacher or an airline stewardess, nothing else." She accepted that advice uncritically.

The Corrigans then sent her to Marian, a private coeducational Catholic high school run by the Boston Archdiocese. She joined almost every club, performed well academically, and in tenth grade met Steve McAuliffe. In a short time they agreed that they would marry, but only after they had both finished college. Four years later, as she worked toward a bachelor's degree in history at Framingham State College, she was shocked by Robert Kennedy's assassination. She watched the war in Vietnam grow uglier, but instead of walking past the protesters to get to class, as Seddon had done at Berkeley, she put on a black armband and joined them.

She admired strong women. Like many young women in 1968, she looked beyond the powerful men who had brought on the chaos, focusing instead on female leaders who demanded change, among them Margaret Thatcher and Indira Gandhi. And in her classes she discovered pioneer diaries, on-the-spot journals written by men and especially women as they crossed the American West in wagon trains in the nineteenth century, often leaving their children in graves alongside the trail.

As they had pledged, Christa and Steve married after they both graduated in 1970. They moved to Washington, D.C., where Steve earned a law degree and Christa a master's in educational administration. In 1978, when Christa was pregnant, they moved to Concord, New Hampshire, where their son Scott was born and Christa found a job teaching at the junior high school. In 1979, a daughter Caroline arrived and Christa stopped teaching until 1982, when a job opened at Concord High School. At last she had a chance to teach social studies the way she had always wanted, with field trips to courts and prisons. She also had a chance to assign those unforgettable pioneer diaries to students in her women's history class.

Teaching is demanding, as is raising two children, and Christa McAuliffe's days were packed with commitments. In addition to her classes at Concord High, she taught Catholic doctrine to children in her parish, led a Girl Scout troop and hosted an inner-city youngster in her home every summer. Rising at 7 A.M., she fed her family breakfast, made their lunches and didn't see home again until well into the evening. Her life was full. But when she heard on the radio that NASA would send a teacher into space, she couldn't get the idea out of her mind. She talked it over with her husband and eventually, at the last minute, she sent for an application. She then worried over it, rereading the questions, trying to figure out how she could make her application stand out. She finally decided to write about the pioneer women who had helped clear the path for their families, and she wrote about her memories of Alan Shepard's flight and about being counseled as a youngster that girls had limited choices when it came to having a career beyond home. She explained in her NASA application:

> As a woman, I have been envious of those men who could participate in the space program and who were encouraged to excel in areas of math and science. I felt that women had indeed been left outside of one of the most exciting careers available. When Sally Ride and other women began to train as astronauts, I could look among my students and see ahead of them an ever-increasing list of opportunities. I cannot join the space program and restart my life as an astronaut, but this opportunity to connect my abilities as an educator with my interests in history and space is a unique opportunity to fulfill my early fantasies. I watched the Space Age being born, and I would like to participate.

She proposed writing a diary as the pioneer women had done. Hers would be a trilogy—the selection process, the flight and the aftermath.

The competition for the teacher's "desk" on the shuttle was fierce. NASA had arranged for the Washington-based Council of Chief State School Officers to form a committee for each state, as well as ones for the Bureau of Indian Affairs and U.S. schools overseas. Each committee was to nominate two candidates. The chair of the New Hampshire committee said that it was nothing like picking the Teacher of the Year. For that they sought a teacher who had successful students who credited that teacher with their success. The New Hampshire chair noted: "NASA wasn't looking for the teacher who communicated the most effectively with students. It wanted the teacher who communicated the most effectively with *everyone.*"

NASA, always conscious of its impact on public opinion, milked the selection process for all the media attention it could win—which was a lot. Eleven thousand teachers applied, but only seventy-nine of the applications came from New Hampshire. The contestants included Rhodes scholars, doctors, a pro football player, several pilots and balloonists and an unsuccessful astronaut candidate. McAuliffe and Robert Veilleux, an astronomy and biology teacher from Manchester, were the winners from New Hampshire. They took medical tests and were videotaped answering three questions. The tapes were then passed on to the judges in Washington. When McAuliffe met with the state selection committee, she had her first television interview. One of the reporters there noticed something special about her: "It was strange, really, because she spoke in perfect, twenty-second bites. She was a natural, but I'm not sure she knew it at the time." She was the right person for this particular spot, not the right teacher necessarily, but the person who resonated best with the media. And this was just the beginning.

When she arrived in Washington she was one of 112 teachers—the picks of fifty states and territories. The teachers enjoyed the all-expense paid visit, complete with workshops, press conferences and a chance to meet astronauts as well as the president. It was great fun, but also stressful. On the fifth day of the visit they would be judged, and on the seventh day the 112 candidates would be winnowed down to ten. The ten would have another three weeks to compete against each other, and then there would be just one. That one would travel into space seven months later, in January 1986.

This selection committee was unlike any other that NASA had put together. It included two college presidents, one from Duke, the other from Vassar, three former astronauts—Deke Slayton, Gene Cernan and Harrison Schmitt—as well as inventor Robert K. Jarvik, who had rocketed to fame by developing an artificial heart. There was also an actress, Pam Dawber, who played the role of a woman involved with an alien (Robin Williams) on the popular television sitcom *Mork and Mindy.* She was there, NASA explained, because she had experienced instant fame, and would be able to explain to the teachers how fame would change their lives. The committee never looked at the candidates' academic records or their teaching experiences.

On the third day in Washington the teachers listened to astronauts Joe Allen and Judy Resnik talk about their experiences in orbit. It was the first time McAuliffe saw Resnik, who openly described her life as an astronaut, including the launch abort on her first mission. Resnik also told the teachers how she had always had motion sickness on amusement park rides and how she had such a fear of heights that the walk high above the ground across the steel-grated bridge to the shuttle was one of the worst moments of the mission. Joe Allen described the astronaut corps as "a very diverse group of people. There is only one thing I can think of that we have in common, and that, with no exception, is that each of us has been fortunate enough to get a very good education, primarily because at some time in our careers we have been associated with excellent teachers." The scientist-astronauts forged a special link with the teacher candidates and even John Glenn, who had just a year before criticized the citizen-in-space program as a poorly conceived public relations ploy, greeted the teachers enthusiastically. The presence of America's favorite astronaut-turned-statesman lent dignity to the lotterylike procedure.

Then it was over. McAuliffe flew home and at three the following morning a telephone call woke her up. She had made the final ten. Eight days later she again left her family to fly with the other nine finalists to Houston, where she was driven south to NASA Road 1 and the Johnson Space Center. The group included men and women, grade school and high school teachers, science and social studies teachers, teachers who had graduated Phi Beta Kappa from top universities and one teacher from a state college.

Despite their varied backgrounds, they bonded into a team with shared concerns and enjoyed the week. One of their worries was the testing process. At lunch in the Center's cafeteria before they were scheduled to go up in the notorious "vomit comet,"* the teachers spotted three astronauts at a nearby table. One of them was Judy Resnik, and she came over to chat. She would be joining them on the vomit comet, she said, and in a whisper advised Christa, woman to woman, not to worry. "Only the men get sick." Christa held up well on that flight as well as in the medical and psychological tests they took afterward and she felt grateful to Judy for her encouragement.

After visiting Houston but before their final interviews back in Washington, the teachers were flown to Huntsville, Alabama, where the Marshall Space Flight Center was celebrating its twenty-fifth anniversary. There was a Space Camp attached to the Center, an educational-entertainment park that offered space travel-like experiences, including a ride in a centrifuge that simulated liftoff. At their arrival, Gregory Walker, a twenty-year-old student-employee, explained to everyone how the ride worked. McAuliffe couldn't make up her mind whether to try it, waffled, declined and then at the last minute walked in and fastened her seat belt. Walker sat near her at a control station. Then the lights went out and the room began to spin faster and faster until it had reached twice the force of gravity. The teachers were listening in the dark to a narrator when they heard voices screaming for an ambulance. Walker had not stayed strapped in. He had fallen and then been hurled through one of the walls into the machinery. When the ride finally stopped, it was Christa who found his injured, unconscious body. Several hours later she learned he had died. That memory haunted her in the next months, even as she denied worrying about the risk on her own flight.

In Washington again, each candidate prepared a forty-five-minute talk for a different committee, all NASA staff this time. NASA wanted to be sure that whoever they chose would be able to handle a variety of pressure from the media, from the training process, from the flight itself and from separation from their families. Accordingly, committee members didn't just sit and listen but instead tried to upset and confuse the candidates. Christa McAuliffe had been captain of her college debate team and she remembered those lessons. Under pressure she paused, then answered calmly. She told them coolly how she would use the experience in orbit to excite kids and teachers about the space program.

---

* A specially modified KC-135, a tanker jet aircraft that can carry many people, and provides about 20 seconds of weightlessness with every loop it makes.

McAuliffe had never been considered the best teacher in Concord, or even an especially popular one. But the television audience is not a classroom, and she displayed an extraordinary ability to communicate on television. She was unpretentious, winning strangers with her ready smile. She was extraordinarily ordinary, and defined herself that way. There are critics of NASA's choice of McAuliffe who argue that "ordinary" is just another way of saying "mediocre," which might have described her intellectual achievements. But "ordinary" does not necessarily mean mediocre. It means rather that she was like most people in the United States who do their jobs, raise families and harbor enormous dreams. NASA did not want a research professor in space—they had their own professors. They wanted a teacher like the teachers most people recalled from childhood—women, for the most part, who made school memorable. The committee chose her unanimously, and no one was surprised, except perhaps Christa herself.

At the same time the committee picked McAuliffe, they chose Barbara Morgan as her backup. An astronaut's backup is like an understudy, ready to go in an emergency. As such, Barbara Morgan shared McAuliffe's life during four months in Houston and she was in Florida for the launch. Morgan is a doctor's daughter from Fresno, a medium-sized city in California's Central Valley. Like Resnik and McAuliffe, she met the man she married while both were still in school, in their case as undergraduates at Stanford University. But they had waited to marry until after she had taught for a year at the Flathead Indian Reservation in Arlee, Montana, and then in Quito, Ecuador. When they married she moved to her husband's home in rural Idaho, where she taught fourth graders. In 1986 Barbara Morgan did not yet have children of her own to leave behind when she moved into a furnished apartment in the same anonymous housing complex as McAuliffe. Morgan accompanied her in every aspect of her training and was a friend as well as a shadow throughout their months in Texas. She is a living diary of that experience.

Morgan's presence is fortunate, because while McAuliffe studied hard in Houston, baked an occasional pie with apples her children had picked in New Hampshire, and worked needlepoint, she did not keep the journal that would have been Part I of her promised trilogy. Maybe she postponed writing when she learned of the unofficial astronaut code of silence, or maybe it was because she had learned about the legal ramifications of the publication, matters of ownership, privacy and publication. We cannot know why she did not write, only that no journal was found among her effects.

Morgan and McAuliffe became close friends. Both were pleased that Judy Resnik would be on Christa's flight—it meant that at least one member of the crew had made her feel that she really wanted her there. Resnik's attitude was a surprise, but then Resnik had never been predictable. She had once told friends that she resented the congressmen payload specialists who were "invading our space," and asked: "What are we going to do with these people?" But a teacher was not a politician. Although she had told friends she did not like the idea of nonscientists taking up valuable time and space on the shuttle, she behaved warmly toward Christa McAuliffe.

Resnik had always been close to her own teachers. She had visited Firestone High School three times after graduating and had invited Donald Nutter, the teacher who had taught her trigonometry and calculus, to her first launch. And when Burton Willeford, who had taught her Algebra II in ninth grade turned up with his family at the Space Center, she autographed a postcard to "my old math teacher." Math teachers, music teachers, science teachers, all had made her accomplishments possible. Flying with the first astronaut of the teacher-in-space program was her way of saying "thank you." The rest of the crew seem to have felt the same way because they all dressed up for a gag photograph wearing mortarboards with tassels, students ready to graduate, in honor of Christa.

McAuliffe appreciated Resnik's help, and was very careful not to trespass on her turf. She was embarrassed when the press reported that she was there to *humanize* the space program, as if the career astronauts were not really human, although she had said those words once herself without considering the implications. That was before she had gotten to know them, Dick Scobee the commander, Mike Smith the pilot, Ron McNair and Ellison Onizuka who were mission specialists like Judy, and Greg Jarvis, a payload specialist like herself—they were special people, all interesting and friendly and very able. She wanted their respect and above all did not want the career astronauts to think she was thumbing a ride. She did not pretend to be a professional astronaut or a scientist—it was no secret that she had had a hard time learning the basic science behind the experiment she was scheduled to conduct in orbit. At one point Commander Scobee had told her "in serious jest" not to touch any of the 1,300 switches in the shuttle cockpit simulator. As for risk, she acknowledged that of course it was present, but she thought of it as the scared feeling you get before riding a roller coaster, the feeling that disappears as soon as the ride begins.

On January 24, at last, the crew of STS-51L strapped themselves into their seats for the launch. Scobee and Smith sat upstairs; Resnik, Onizuka and McNair sat behind them, and the two payload specialists, McAuliffe and Jarvis, sat on the level below. They waited patiently, but as often happens with space shuttles, the launch was scrubbed. So were the next two. The crew kept eating farewell dinners, until finally, on the morning of January 28, after having been held back another day because of 30-mile-an-hour winds at the launch site, they were ready to go. The evening before, the temperature had kept dropping, and throughout the night, NASA officials talked to each other at different NASA centers and to the engineers at Morton Thiokol in Utah who had built the rockets. Was it too cold to fly? The engineers were worried. It had never been this cold at a launch. As the crew slept, the mercury dropped to 27 degrees and at the launch pad icicles two feet long hung from the upper level of the spacecraft. The cold froze an oxygen sensor on the shuttle as well as a pair of video cameras on the ground. A coating of ice embraced the shuttle.

At 6 A.M. the crew awoke and met for breakfast. The table was covered with American flags and a cake decorated with the *Challenger* crew patch: the stars and stripes behind a globe of the Western Hemisphere and the names of crew members in groups, the mission specialists on top, the commander and pilot below, and below them the payload specialists. They ate steak and eggs. Judy Resnik, always hungry, had two helpings. Then they changed into their NASA blue jumpsuits and left for the pad.

At 8:15 they put on their helmets and flight vests. The vests carried small inflatable pillows in the back to ease the discomfort of lying on the hard seats.

As they climbed into the crew cabin, a quality assurance inspector gave McAuliffe a red apple "for the teacher." She thanked him and asked him to save it. "I'll eat it when I get back."

Then they waited. It was cold inside the cabin and the crew complained about the temperature and about having to lie still for so long. Resnik said her butt was "dead already."

At 10 A.M. the temperature had risen to the freezing point, 32 degrees Fahrenheit. An hour later the temperature had risen to 37 and Launch Control told Scobee: "We're planning to come out of this hold on time."

Students in schoolrooms around the country watched CNN.

Scobee told his crew: "Seven minutes!"

Resnik shouted "All right!"

Marvin Resnik remembers: "When the *Challenger* went up, we were in the waiting room, the ready room. And Steve McAuliffe and I were just talking. He said, 'Marv, I'm scared to death.' It's the only time I ever heard anyone say anything about fear. I said, 'But none of them [the astronauts] are. You don't really have anything to worry about.' Famous last words." He remembered how Judy had reassured him. "'Daddy, I know that I'm sitting on a bomb, but NASA's got everything figured. They have a redundancy for everything'—Except what happened."

What happened was seen firsthand by friends and family members in the windy grandstand facing Banana Creek, and by workers on the Florida freeways who stopped what they were doing to look up, never tired of the spectacle.

Launches are oddly silent at first. The rocket rises straight up and leaves a plume of smoke. Then, after what seems like a very long time, a boom resonates that sounds, according to Martin Harwit, former director of the National Air and Space Museum who happened to have been a sailor on a ship near the Bikini atoll in 1949, "just like the H-bomb." Seconds later there is a shock wave. That's what NASA regulars expect.

But this launch was different. As thousands watched, seventy-three seconds into the flight the smoke trail broke into a zigzag and the shuttle disappeared. Those watching did not understand immediately, or want to understand. But the *Challenger* crew knew.[†]

"Roger, go at throttle up," Commander Scobee's voice echoed in the control rooms in Florida and Texas just moments before an explosion ripped through the orbiter. It was later discovered that two members of the *Challenger* crew had turned on the valves in their air packs, meaning that they were still conscious and expected to lose oxygen. The initial blast had probably blown the crew compartment away from the forward fuselage, the cargo bay, the nose cone and the forward reaction control compartment, but it had not shattered the walls of the compartment or injured the crew. All seven astronauts were strapped in their seats when, less than three minutes later, their intact cabin smashed into the water, shattering instantly upon impact. The physician who performed the autopsies believed they were unconscious within seconds of the explosion. At the Space Center, no one alerted the nearby hospital in Titusville. They knew it was hopeless.

The crew's families were brought together quickly. Other astronauts joined them, along with Barbara Morgan, Christa's back-up. She remembers how brave the families were, even in their intense grief. And when the vice-president, George H. W. Bush, came to speak to them, they told him how important it was to them, in the spirit of their lost husbands, wives, children, brothers and sisters, to continue human space flight. They agreed that they did not want a brick-and-mortar monument, but rather a way to perpetuate the spirit of the lost mission, to educate children especially about space flight. That decided, the families joined to found the Challenger Center for Space Science Education.

NASA got the Resnik family to the airport in Orlando, where they were put on a flight home under assumed names. In an effort to console her step-

---

[†]A cockpit tape found after the accident has Mike Smith saying, "Uh-oh!"

father, Judy's stepsister and close friend reminded him, "She wanted to do this." They had spoken the day before, and Judy had been eager to get going.

The crew of STS-61D, the much-delayed science mission, watched in horror. Millie Hughes-Fulford, a payload specialist in the group, recalled: "Actually *we* were scheduled for pad 39B. Our flight was scheduled, but then that [animal subject] cage failed a test. *Challenger* went forward instead of us. We had a meeting right before the flight and people were joking around saying, 'There goes your flight. A schoolteacher took your place.' At NASA if you're in a meeting, there's a closed circuit TV in most rooms. You always stop and watch a launch. We were watching it. You could tell it was gone. The meeting just broke apart. No one said 'we're breaking up.' People just wandered off. It was really a shock."

Rhea Seddon, who was also in the Spacelab crew, recalls, "We were watching that launch and said to ourselves, 'We should have been on that one.'" She continues: "It was a horrible thing to go through, and to imagine what it would have been like and to see what happened to the families, and to understand why it happened. I think it was pretty devastating for all of us. For me personally, I'd never lost any friends. . . . I had never had to deal with grieving families, grieving friends. I had never had it come so close to me." Later she was drafted into helping reassemble *Challenger,* as pieces were salvaged from beneath 73 feet of water, inside a giant hangar at the Kennedy Space Center.

But almost as disturbing as the task of reassembling *Challenger* was the way NASA handled the grieving process. According to Seddon, the men really denied the fact that this was a tremendously stressful time. "In working with some of the flight surgeons I knew that the spouses were going through very difficult times. And I knew that a number of the women . . . felt that the flight surgeons ought to provide some psychological support, or offer to. That did not go anywhere with the astronaut office. They said, 'We don't need it. We're doing fine.' To me it said that the male thing had been, 'We will just deny that there is anything wrong. . . . We don't want anybody climbing inside our heads and we don't want to admit to anybody that we're hurt.'"

Then there was all the uncertainty. As Seddon recalls, "Everyone [was] calling into question whether or not we should continue with the space program. I had made a pretty large investment by that time. . . . It was kind of a question of 'Is this job going to exist next year?'. . . It was a pretty terrible year."

A week after the disaster President Reagan created a commission headed by former attorney general William P. Rogers to find out what had happened. All the publicity surrounding the tragedy, including the loss of Christa McAuliffe, seemed to warrant presidential action. At NASA, especially among those who remembered the Apollo tragedy of 1967, when three astronauts died tragically in a test on the launch pad, many people were extremely disappointed that NASA was not conducting its own investigation. They were afraid that the public would sense that NASA had lost its competence and perhaps its self-confidence. They also worried that an investigation carried out on television would be politically driven and would end up, as a former NASA center director explains, "trying to pin the blame on someone for what had happened rather than seek out the root cause of the disaster." This indeed did occur, not through any fault of the commission, but because in the agony of national mourning the public wanted (and perhaps needed) to find someone or some one thing responsible.

As it was, *Challenger* happened at a time of unprecedented confusion at NASA headquarters in Washington, where a new administrator had yet to win the support of his staff, and was not in touch with his Center directors. For the only time in its history, neither of NASA's two most senior people, the administrator and the deputy administrator, had come to the launch. In the circumstances, the agency had to accept without protest the demand for a public investigation. NASA never did initiate its own.

The Rogers Commission had an all-star cast including Richard Feynman, the eccentric Nobel Laureate physicist from the California Institute of Technology, and astronauts Sally Ride and Neil Armstrong, as well as many senior industry and military representatives. By July, 120 days after it began, it had finished its deliberations. There was no difficulty pinpointing the O-rings as the technological weakness that had allowed the 10,000°F gases to leak out during ignition of the solid rocket motor. To emphasize that finding, Feynman dropped a miniature O-ring into a glass of ice water before television cameras, demonstrating its lack of flexibility in low temperatures. To most engineers, this was no surprise. NASA had been aware of the O-ring problem since the midseventies, when references to them first appear in the files. The O-rings were designed to seal a tiny gap, a fraction of an inch, in the joints of the solid rocket motors. The design included two O-rings, a built-in redundancy so that if the first should fail, the second would block any flames from spreading. But redundancy did not help. In fact there were three separate failures that, together, led to the disaster. First was the rigidity of the O-rings in the unusually cold temperature. Second was the imperfect fit of the solid rocket motor (SRM) segments because of damage during a prior launch that had left them not perfectly round. Third was the shuttle's encounter with high-altitude wind shear conditions early in its ascent. This created huge forces on the orbiter that caused the steering system to be more active than on any previous flight and put high dynamic loads on the SRM field joints. The joints failed, and the flames spread until the entire shuttle exploded.

The Rogers Commission concluded: "The failure was due to a faulty design unacceptably sensitive to a number of factors. These factors were the effects of temperature—physical dimensions, the character of materials, the effects of reusability, processing and the reaction of the joint to dynamic loading." The lack of resilience of the O-rings, so dramatically demonstrated on television, was not the sole fault, and the reliance on redundancy, which had made Resnik so confident, had not been sufficient to protect against a catastrophe.

The commission also concluded that the decision process that led to the launching of *Challenger* was "flawed." They pointed to the long history of concerns with the O-rings by engineers at Morton Thiokol, the company that made the rockets—concerns that were never brought to the attention of the people making the launch decision. They found a closed culture at NASA and its major contractors that refused to listen to warnings or criticism from those engineers who tried to warn them. In the end, *Challenger* is a nightmare example of Murphy's Law. The weather, bad design decisions, poor engineering implementation and an inadequate decision-making process colluded to allow the unimaginable to happen.

NASA accepted these conclusions and during the next four years set about redesigning the hardware and also its decision-making procedures. The flaws found by the Rogers Commission never have been second-guessed and were assumed to have been corrected.

The public at large found the temperature-related O-ring weakness easy to understand and blamed NASA's managerial incompetence for the disaster. Most of NASA's senior people were replaced, losing the wise as well as the incompetent in the process. Wherever the faults lay, the magic days of Apollo were over for good, and with them NASA's perception of itself as "one of the primary instruments of the Nation's cold war arsenal." With its aura of invulnerability gone, NASA would be more than ever beholden to public opinion. And the public, which NASA had bathed in pre-launch sound bites about its first citizen-astronaut, now followed every excruciating detail of the disaster.

As shuttle debris floated up to the surface, McAuliffe's lesson plans were retrieved by rescue teams. Her body and the remains of the other astronauts were removed in March from the crew compartment on the seafloor. Scobee and Smith were buried in Arlington National Cemetery. The remains of the others went to their respective home states. Judy Resnik is buried in a Jewish cemetery in Akron, Ohio. Christa McAuliffe is in a Catholic cemetery in Concord, New Hampshire. The financial settlements that were reached by the families of the lost astronauts are private. McAuliffe's parents, the Corrigans, respected the astronaut code of silence until after Ed Corrigan's death in 1990. Only then did Christa's mother, Grace, write a memoir that included an unsent letter from her late husband that she had found among his papers. It said, in part:

> My daughter Christa McAuliffe was not an astronaut—she did not die for NASA and the space program—she died because of NASA and its egos, marginal decisions, ignorance, and irresponsibility. . . . One of the Commissioners stated, "It was no accident; it was a mistake that was allowed to happen."
>
> President Reagan said that the act was not deliberate, was not criminal. But I say that the sins of omission are no less sins than those of commission. I say "they" deliberately neglected to make necessary corrections to the O-rings and are, therefore, as guilty as if they planned a deliberate criminal act. My attitude, I am sure, differs from that of astronauts' spouses and families. I feel no allegiance to NASA.

His feelings were perhaps *not* so different from those of the astronaut families. They all filed lawsuits against NASA and Morton Thiokol. The McAuliffes along with three other families accepted some $7.7 million dollars in December 1986, the sums based on the age of the deceased and the number of dependents. The Resniks and the McNairs settled independently over a year later. The last family to settle, the widow of Mike Smith, filed a $1.5 billion claim against Morton Thiokol explaining: "No one in big business should be allowed to make a faulty product and profit from it."

In 1987, a reporter asked Sally Ride if she was ready to go up in the shuttle again. She responded with an unqualified "No," not as things were then. But she remained committed to NASA and to human space flight. The *Challenger* disaster did not, in fact, erode the confidence of many of the astronauts. None of the women quit. They had known it was a risky business from the start, and felt that, if anything, the soul searching and investigation of what had gone wrong could only result in greater safety for those who remained in the program.

In August 1987, Ride presented her legacy to the agency in the form of a report entitled "Leadership and America's Future in Space." The Ride Report, as it became known, was a product of the commission she headed whose directive was to clarify NASA's mission in terms of both robotic and human space flight. The report suggested several specific goals for NASA, including exploration of the solar system, the establishment of an outpost on the Moon and a plan to send humans to Mars. The report urged the return of the shuttle to flight status, and recommended that the United States use the technologies learned from space exploration to improve the quality of life on Earth.

A few months later, Ride resigned from NASA. She had flown two shuttle missions, contributed to the Rogers Commission and chaired her own. NASA expected its astronauts to remain in the corps for at least six years. Ride left after eight. She never publicly blamed NASA for the tragedy and explained her departure as a longing to rejoin the academic world, and accepted a place at Stanford. She was the first American woman to go into space, and the first to resign from the astronaut corps.

Challenger's tenth trip, NASA's twenty-fifth shuttle launch, had received extraordinary attention from the moment NASA announced it would carry a teacher. Television cameras followed Christa's odyssey from her selection through the 73 seconds of flight. The media also covered what they could of the lives of the rest of the crew. With their names and faces broadcast around the world, NASA showed that the astronaut program embraced different races, religions and genders. NASA had made good on its promise to treat its new mission specialists—women, men and people of color—equally.

But the decision to fly a "citizen-astronaut" may have been a terrible miscalculation. Launching a woman who was a teacher and a mother of young children made it in some ways a poster child of poor judgment. As NASA sought to retrieve the remains, the world was reminded that two women had died with their five male crewmates. Resnik's martyrdom was of a piece with that of the rest of the crew. As professional astronauts, they were aware of the risk. McAuliffe's death was, of course, a public relations nightmare, as well as a personally felt tragedy. McAuliffe's embodiment of feminine innocence and perky curiosity reminded the American public that the space shuttle was still an experimental craft and that it was too soon to launch just anyone into orbit.

NASA's reputation had been tarnished. Although the shuttle had been behind schedule and over budget from the start, its supporters in Congress and the public had respected NASA's management and believed its promises. NASA had seemed to have a magic touch as exemplified by the quasi-mythic retelling of the saga in the movie *Apollo 13* of the accident that disabled the spaceship, and the subsequent rescue that had enhanced NASA's "can do" reputation. *Challenger* destroyed that aura of invincibility, probably permanently. In its newly vulnerable state, NASA's leaders knew that the media was watching as never before, and that they were more than ever before beholden to public approval.

7

FROM THE ASHES

THE *CHALLENGER* DISASTER brought the human space program to an abrupt halt. Astronauts and their support teams were forced to adjust to seemingly endless delays and changing procedures. In 1986 five women remained from the class of 1978 as well as two from the class of 1980. NASA had accepted two more women as mission specialists in the class of 1984 and another two in 1985. Three other women had been training as payload specialists at the time of the accident under special agreements. Two of the three were foreign nationals. They were all highly skilled life scientists who had been assigned to missions scheduled for later that year. The uncertain length of the delay was hard on every astronaut, but for women listening to the ticking of their biological clocks, the stress was probably worse.

While the astronauts waited, NASA administrators set about reexamining the shuttle's purpose. They had been under pressure even before *Challenger* to demonstrate that the shuttle—up and running for four years—was on the verge of fulfilling its promise: that it could keep to a predictable and frequent schedule, that space travel was safe, even routine. President Reagan recognized the place of the astronaut program in the American psyche and he made it clear right away that shuttles would continue to fly. They would fly again as a memorial to the martyred heroes. Reborn, the shuttle would show the world that Americans do not give up. The *Apollo 1* tragedy had led to improvements that sped American men to the moon. *Challenger*'s tragedy would transform the shuttle program.

Yet the shuttle was a different kind of mission in a different political climate. During the sixties the race to the Moon had a clear national goal. A generation later that race had been won and competition in space had lost its edge with both contenders. When Svetlana Savitskaia was selected to fly in

1982, the decision had been made in Energia, not the Kremlin. Recall that, likewise, no NASA administrator had bothered to come to *Challenger's* launch. Competition between the Cold War giants continued, of course, but space was no longer the arena and the United States and Russia were even beginning to see space as a place for international cooperation.

In the shadow that fell across the shuttle program a vocal minority began to ask whether it was worth the risk to send people into space at all. The contrast between *Challenger's* catastrophic failure and the smashing success of NASA's unmanned missions was hard to ignore. These missions appealed to a different constituency, of course, but they nonetheless dazzled the world with their discoveries. Throughout the late seventies and early eighties, staff at the Jet Propulsion Laboratory celebrated as *Voyager* passed by one planet after another, sending back mind-bending, color images of Jupiter's moons, Jupiter itself and Saturn. News of the *Challenger* tragedy reached JPL as the staff was celebrating *Voyager's* flyby of distant Uranus. Inevitably Congress and the public could only wonder: Why risk lives when we can learn so much from an unmanned craft?

NASA has always had to justify the expense and risk of human space exploration. If NASA's purpose was purely exploration, then robots and telescopes were undoubtedly better. But people are fascinated by the way other people respond to the strangeness of space, and many people, not simply trekkies, want to know about the Moon and Mars—and they want to know it from a personal perspective. Why astronauts? The astronaut Rhea Seddon recalls a cartoon she saw shortly after the accident. "There was a motorized Conestoga wagon with a caption that said 'This unmanned thing was sent out to explore the west?'" The shuttle wasn't exactly a Conestoga wagon, but as Seddon puts it, "There were so many things people could do that robots couldn't."

Just as *Challenger* provoked one constituency to question the value of human spaceflight, it prompted others to reopen the old debate over whether to bring women along. Some of the old NASA hands who had doubted the wisdom of accepting women as mission specialists were convinced that they had been right. The Mercury Seven and most of the Apollo astronauts had been test pilots, people familiar with the risk of death. They wondered if Judy Resnik, or the other women in the program, had been truly aware of the danger. Like John Glenn in his 1962 testimony to the congressional subcommittee, these old hands felt that traditionally, perhaps even biologically, men were programmed to protect women, not expose them to danger.

No one seriously considered abandoning the astronaut program or removing women from it, but everyone at NASA knew change was inevitable. The astronauts were especially anxious. They kept training, but business was not usual. NASA was still figuring out which missions would fly, in what order, and with which astronauts, and when. Those not quite *in,* like the payload specialists, kept hoping that they were still part of the program. The mission specialists assumed they would fly and still they worried. They had no sense of where the shuttle program stood in the American imagination.

But they were about to find out. In December 1985 the National Science Foundation had hired the Public Opinion Laboratory at Northern Illinois University to sample the opinions of some 2,000 Americans on a range of subjects, including space exploration. This was one of the first computer-based telephone polls, which meant that it was simple for the pollsters to re-interview the same people in January, just three days after the accident, and again a third time after the release of the Rogers Commission report.

The poll revealed that at the end of 1985 the steady ten-year decline in public interest in the space program had stopped, a decline that had bottomed out with only 45 percent favoring the program. Perhaps the publicity surrounding the teacher-in-space had made the difference, because in late December—five weeks before McAuliffe was scheduled to fly—a majority of Americans had agreed for the first time in a decade that the benefit of the space program exceeded its cost. In November before the accident, about half of those polled recognized Christa McAuliffe's name. Eighteen hours after the accident her name recognition had leaped to 98 percent.

Almost everyone had seen the disaster live on television, or in replays. Psychologists likened it to the national trauma of the Kennedy assassination. Here, as with the assassination, most Americans felt a personal loss, like a death in their families. The nation embraced the memories of all the dead astronauts, but especially the schoolteacher. They dismissed the explosion as a "minor setback" and wanted astronauts back in orbit as soon as possible. According to a poll taken five months after the accident and after NASA had acknowledged it would be a long time before any shuttle could fly, the public still felt a "strong—almost patriotic—loyalty to the space program." The National Science Foundation concluded that the net effect of the disaster was a strong shift in public sentiment in favor of the space program in general, and the space shuttle in particular. The NSF praised NASA's handling of the tragedy, especially its candid review of the causes. The public relations people could only conclude that letting it all hang out had restored, rather than weakened, public support.

This support translated into money. Not counting the one-time $2.5 billion appropriation to replace the lost shuttle, NASA's budget doubled in a period of a few years from $7 billion to over $14 billion. The shuttle would fly again. And four of its future female passengers would make a special mark on a fragile post-*Challenger* era.

After the very public death of Christa McAuliffe, NASA began to reexamine the use of payload specialists. There was no official policy change, and those with crew assignments kept working and waiting along with everyone else. However, as months stretched into years, it became clear that notwithstanding the promise to Barbara Morgan, there would never be another payload specialist teacher-in-space.* Still, three of the women astronauts assigned to flights were payload specialists, of whom two were from partner agencies. They had all been training with their crews and keeping up with their scientific research as they awaited the call to fly. They were guaranteed their flights.

The first woman payload specialist with an assigned mission was Millie Hughes-Fulford, a biochemist at the Veterans Administration Hospital in San Francisco. In 1976 (as Millie Riley) she had applied to the astronaut corps and was rejected at the penultimate cut. Now she was at the Johnson Space Center in a different capacity, on what turned out to be a very long-term loan. A contemporary of Shannon Lucid, she is a tall, gregarious, fifth-generation Texan who grew up oblivious to the gender biases that permeated the world beyond her home in Mineral Wells, a town several hundred miles south of Dallas. She grew up in the fifties with parents who encouraged her to do whatever she liked. One of the things she liked was watching television, especially *Buck Rogers* and *Rocky Jones, Space Cadet.* About then, at five, she decided to become an astronaut. When she was eleven, her father, crippled with arthritis, drafted Millie to work alongside a male cousin in his grocery store. At the store she climbed ladders and carried a "carton of Green Giant peas." No less was expected of her than of her cousin, and she delivered what was expected.

At sixteen, she still wanted to be an astronaut, but on her birthday in 1962, she "realized that all the astronauts were men" and that they were all

*After the accident Morgan had volunteered to speak to many schools for NASA about the importance of the space program in education. She had asked to fly quickly, to "get back on the horse," to show America's children how to fight fear. Eventually, thirteen years later in the flush of attention surrounding John Glenn's second flight, NASA admitted Morgan as an astronaut candidate in the class of 2000. She had been assigned to fly in 2004.

shorter, too. After all, they had to fit into a *Mercury* capsule. At that time, she recalls, she had never heard of the idea of a "glass ceiling."

> [I didn't know] that there were things women couldn't do, because I was getting those Green Giant peas down. No one told me I couldn't do it. I was changing oil in the car. I was doing everything the boys were doing. I didn't see that there were limits. It wasn't until I was sixteen that it occurred to me that perhaps women were not treated the same. And at that point I said, "Fine. I'll be a scientist." I started in psychology and I realized that all psychology was biology and [then] I realized all biology was chemistry. . . . I went into strict chemistry and then I went into biochemistry.

Hughes-Fulford entered college at sixteen, married the boy in front of her in her biology class, had her daughter at nineteen, graduated and went on to graduate school, divorced, and found the job at the VA in San Francisco. After failing to be accepted by NASA in 1976, she continued working at the hospital and kept reading about space exploration. In her enthusiasm, she remarked to a senior colleague that there was a problem with bone loss in space. He then applied to NASA to do an experiment "on the first medical mission." His proposal was accepted and he called her with the news, adding that he could nominate two people to go along. Would she like to go? She said yes. It was 1983 and her research carried a complicated schedule using some laboratory animals and some human volunteers.

Recently remarried, she moved to Houston with her daughter and husband, a pilot for United Airlines. They expected to be Texans for about two years. The mission, Spacelab 4, had already been delayed several times before *Challenger* because it was more difficult than NASA had realized to construct cages for the rats and jellyfish, test animals whose responses to weightlessness would add perspective to what was known about humans. Without the disaster, the mission would probably have flown in 1988 or 1989. With the added delay, Hughes-Fulford did not know how long she would have to wait. She was in an odd, but not uncomfortable position—on loan from one government agency to another.

As it was, she had plenty to do in Houston. There were twenty-eight different experiments and the entire crew had to be cross-trained, in case someone got deathly ill and another had to fill in. It was not easy. Equipment, some left over from *Skylab* in 1974, would break and have to be redesigned. That took up most of Hughes-Fulford's daytime hours, and almost every night she would go into the labs at the Johnson Space Center and do her own science. In 1987 her family returned to California but she stayed. They barely saw her anyway—she was too busy perfecting a routine for a flight that was not quite scheduled.

The first Spacelab Science Mission (SLS-1) was finally ready to go in June 1991 on STS-40, the shuttle *Columbia.* This long-planned mission was the first dedicated entirely to studying the physiological effects of spaceflight on the seven crewmembers—all of whom had undergone baseline testing on Earth for comparison. There were three women: Rhea Seddon, the physician who had already flown in 1984; Tammy Jernigan, a rookie astrophysicist who had joined in the class of 1985; and Hughes-Fulford, a payload specialist with expertise in the biochemical depletion of calcium and other minerals in mammalian bones. NASA already knew that exposure to microgravity led to difficulty retaining balance, loss of calcium from bones and loss of muscle mass and protein. Hughes-Fulford was there to do the experiments on calcium loss. There were eighteen experiments planned, of which ten used crewmembers as subjects, seven used laboratory rats—all male—and one used jellyfish. This was the first major physiological study since *Skylab,* fifteen years earlier, and the crew voluntarily followed demanding pre- and post-flight tests, including twenty-four hours of head-down bed rest on Earth to test whether this ground-based experience was akin to the results found in space.

In June 1991, eight and a half years after taking a leave from her job in San Francisco, Hughes-Fulford's family flew to Florida to watch the launch of STS-40. It was the sixteenth launch since *Challenger* and that memory, still fresh, overwhelmed her daughter, who kept crying "You're going to die!" Hughes-Fulford responded, "I'm not going to die." But of course there was risk, she later reflected. "Any time you climb in a rocket ship you're in danger. [What's] interesting is I did not know how I would feel the day that we were getting ready to take off. I didn't know if I'd be nervous. You don't know. You haven't been there before. I felt that if something did happen I'd have a school named after me, or a road named after me. And that would be OK. I was doing what I wanted to do. It's easier when you're climbing on the ship yourself. But your family . . . It's really tough on the family."

The family waited through two scrubs and then watched the ship go up. Hughes-Fulford remembers, "It was great. The sheer shaking was incredible. No one told me about the vibrations. No one told me about the noise. It sounded like something exploded." And of course something had: the rockets. Eight minutes later they were in orbit and it was time to change clothes and remove the partial pressure suits, something like those worn by their predecessors in the Apollo program, which had been eliminated with the earlier shuttle flights but had become routine again since *Challenger.* The crew took off their suits one at a time, starting with the two payload specialists, Hughes-Fulford and Drew Gaffney, a cardiologist and expert in space physiology. Gaffney was a co-investigator on an experiment studying human cardiovascular adaptation to weightlessness. It took forty-five minutes for the entire crew to doff helmets and suits. Hughes-Fulford explained, "You can't all do it [at] the same time. The shuttle is pretty small."

She knew the flight plan, but she could not start work immediately. "I was so sick. I was the first and I was sick until I took Phernegan [an anti-nausea drug], I think it was the fluid shift, the loss of gravity. I started feeling sick instantly." She was suffering Space Adaptation Syndrome, a discomfort that about 80 percent of space travelers experience (but few admit to) and that can be costly on a tightly scheduled shuttle mission, where lost time cannot be made up.

After a few hours she was able to work on the experiments she had been training to do for most of a decade—medical experiments that used the caged rats as models, as well as the rest of the crew, herself included. She explained, "Normally, if there's any medical part, the crew is the guinea pig. Who else are you going to test? All medical data from flights come from the crew." This was the first all-life-science research flight, and Hughes-Fulford may not have been aware that many crewmembers in earlier flights had declined to be "guinea pigs," and would, in future flights, retain the privilege of opting out of experiments. In fact STS-40 collected more information about the response of the human body to weightlessness than all of the research done on previous missions. It would take several years to analyze all the data they collected on the human skeleton, heart and lungs. One thing that the medical researchers did not look for, however, was any difference in physiological reactions between women and men.

Hughes-Fulford worked steadily, amassing data about calcium loss. The problem of bone demineralization has ramifications apart from the problems of space travel. As the population ages, osteoporosis, a condition resulting from calcium loss, afflicts larger numbers of people. NASA explained the decision to send seventy-seven-year-old John Glenn back into orbit in 1999 as an opportunity to study the problem in an older man.

Back in her office at the VA Hospital in San Francisco, Hughes-Fulford wonders about bone loss in the aging population and, remembering the freedom that weightlessness afforded her movements in orbit, she ponders the advantages that weightlessness could provide paraplegics. A cartoon posted on her office door at the VA Hospital shows a man leaning on a walker in the first box, and in the next box he is using the walker to do a handstand. The caption reads, "I blame John Glenn for this." In Miami, Hughes-Fulford notes, the VA runs a program where paraplegics scuba dive and "to a person they all say it is the only freedom from gravity they have ever known." How much more freedom they would have, she thinks, in a distant future where seriously paralyzed patients could visit "space hotels" where they could move about in microgravity.

Hughes-Fulford was lucky to fly when she did, even if she had to wait a long time. She turned out to be the only American woman, besides McAuliffe, to fly as a payload specialist. NASA did not eliminate the position entirely, but in the general reorganization of the late 1980s, with a few exceptions (one being John Glenn's second flight), NASA drastically curtailed its use of American payload specialists. There was never an explicit explanation of the changed policy, but it was probably a way of using their growing reserve of mission specialists, many highly skilled in the kind of work the newer missions demanded.

That said, NASA continued using payload specialists from its partner agencies in other countries. Before *Challenger,* when NASA expected to operate on a predictable schedule, the agency had established partnerships with space agencies in Canada and Japan. NASA promised to train and carry their astronauts as payload specialists in exchange for money and technological help. In accordance, both Canada and Japan had selected six and three astronauts, respectively. Whether the American women's movement had made these democracies sensitive to their female constituents, or if women were among the best candidates, one of each country's choices was a woman. Each nation flew a male astronaut first, but only Canada's Marc Garneau flew before *Challenger.* The women, both physicians, and the Japanese male physicist were caught in the same holding pattern as the rest of the American astronaut corps.

The wait may have been hardest on Roberta Bondar, a neurologist from Sault Sainte Marie, Ontario, and the only woman among the six Canadian semifinalists and one of two astronauts selected in 1983. Bondar had what can only be called a prickly personality, a trait that did not limit her career in medical research but that handicapped her with Canadian reporters, who were generally more aggressive than their American counterparts. Bondar also didn't have the benefit of NASA's tutorial on how to handle the personal questions about her love life, or lack of love life, that her American colleagues were spared.

At first, when she arrived in Houston in 1984, the Canadian press simply reported that she had begun training as prime payload specialist for the first International Microgravity Laboratory on the *Columbia,* STS-71-I. She would perform life science experiments on the mission as part of the partnership agreement. Bondar, like other women born in the fifties, grew up dreaming of space travel, only to realize as she grew older that her dream was twice blocked—once by gender and again by nationality. Canada had no space program. So she continued dreaming of space but became a physician.

Canada may share a border and a language with the United States, but it is a country with its own history, politics and outlook on the world. Like the United States, Canada has welcomed immigrants, and as in the United States, newcomers have not always found it easy to fit in. Bondar's father Edmund emigrated from the Ukraine and her mother Elaine came to Canada from Wales. They felt like outsiders and were determined that their daughters—Roberta and an older sister, both born in Canada—would be accepted. But despite Brownies, Girl Guides and the YMCA, their younger daughter was different. She was an athlete when girls weren't admired for athletic prowess, and she did not wear makeup or fuss with her clothes in high school. Instead, she focused on science.

Her family encouraged her interest. She recalls in her account of her mission how at five she convinced her mother to send a coupon to the Double Bubble Company for a "space helmet." When the cardboard helmet arrived, five-year-old Bondar put it on and set out to explore nearby McGregor Avenue. Later, in seventh grade, her uncle, a pharmacist, helped her set up a laboratory in her basement with some of his leftover test tubes. By high school she excelled in science and had joined the math and science clubs.

She never lost her interest in space. She eventually bought a model rocket kit and watched every NASA launch covered by Canadian television. Meanwhile, she went on to college and earned first a bachelor of science degree in agriculture and zoology and then a Ph.D. in neurobiology and an M.D. specializing in neurology. Along the way she got a pilot's license, ready for whatever could happen.

What happened was that in 1983 Bondar learned that the Canadian Space Agency had arranged with NASA to send three Canadians as payload specialists on shuttle flights. Along with 4,300 other Canadians, she applied, passing through finer and finer sieves until the new Canadian Space Agency announced its choices. There was pressure then in Canada, as there had been in the United States, to include a woman as well as someone from French-speaking Quebec. When Bondar's selection was announced along with the others, one member of the committee reportedly remarked that they were relieved to have found her, which meant they didn't "have to force the system," to bend the rules to select a woman. Bondar was at least as well qualified as the other winners.

They arrived in Houston together in 1984. NASA had promised the Canadians two shuttle missions, one in 1985 and the other in 1986. Garneau flew in 1985, on schedule. As Bondar awaited her flight she was shaken in early January 1986 to learn of her father's sudden death, followed almost immediately by the *Challenger* accident. As she attended services in Florida for the lost astronauts she noted their parents' grief. She pondered the risk of inflicting similar sorrow on her own close-knit family. She stayed, of course. She had her assignment. Bondar was the only woman in the crew, but there was another payload specialist, Ulf Merbold, from the European Space Agency. They would investigate the effects of microgravity on the human nervous system, Bondar's specialty, and on other life-forms.

Like Hughes-Fulford, Bondar waited for her mission, assuming at first that the delay meant only a few months, then a year or two at most. She continued to train in Houston, although after a while she began commuting to Toronto, where she had left her patients—stroke victims at the Sunnybrook Health Science Centre and Toronto General Hospital. She returned to them, and to her research, using a new kind of transcranial doppler ultrasound to measure blood flow in the brain. Later, in space, she would use a similar instrument to measure the crew's cerebral blood flow to see the effect of fluid shifts in microgravity. This would help the medical group track headaches as well as any extraordinary conditions brought on by the rise of liquids to the head, a major consequence of living in microgravity.

The years stretched on and Bondar continued commuting, spending weeks and months at a time in Houston preparing for the mission, and the rest of the time at the Canadian hospital with her patients. During these years, she put her social life on hold. Always shy in social situations, she had no time to meet a partner.

Then the waiting was over. Nine years after she had arrived in Houston and after nineteen postponements, her flight was scheduled. Sixty-five family members and friends came from Sault Sainte Marie in January 1992. In Canada, her hometown opened a Winter Park in her name and commissioned a sculptor to carve her image and a life-size replica of the shuttle *Discovery* from eighty tons of ice. Canada went wild over Bondar. She responded by singing the Canadian anthem, on and off, throughout the eight-day flight.

In orbit she was either studying the way the rest of the crew functioned without gravity or she was in the lab studying biological samples of fruit flies, frogs' eggs and slime mold. The scientists in the crew worked a split shift to utilize every minute of laboratory time. They took turns sleeping, sharing bunks, but they were an odd number and Bondar got a bunk to herself, not because she was female, but because she was shortest and could squeeze into a nook where no one else fit.

Like most astronauts, she marveled at the beauty of the stars visible in orbit, but enjoyed an unanticipated bonus. She had entered the shuttle wearing glasses, only to discover that she didn't need them in orbit. Apparently, she learned, the fluid shift that distorts vision in some astronauts corrects vision in others.

In 1994 Bondar published *Touching the Earth,* a memoir and photo-essay of her mission. She describes the joy of finally being in orbit, of gliding along a ladder to the flight deck without having to take a step. She longed for an instant of time travel, just enough time to tell "the little girl on McGregor Street" in the cardboard space helmet that one day she would make it into space. In her book she refers to "a lifetime of coping with issues of role and gender" and she apparently took pleasure in seeing them vanish as she was "transported back to the innocence, energy and imagination of my youth." Bondar continues, "These childhood qualities never really left me, but I certainly have been bruised by struggling upstream on Earth, following my dream against all odds."

In Canada, one of six astronauts at the time, the press dogged her steps. They noticed, of course, that their only female astronaut was single, and they plagued her with questions about the children she did not have and the life she had not led. Finally, cornered by them, she confessed that when she had become an astronaut in 1983 she had still thought of having family. But the wait had been longer than she had expected, and during those years she had been too busy to have a social life. Then it was too late. None of NASA's American women were harassed this way, although many female astronauts had no children and no spouses either. Neither did some of the men. Those who did not like being in the spotlight because they wanted a private life, like Judy Resnik, deflected these questions with a cool shake of the head. Bondar did not have this skill. Her friends described her as "not suffering fools kindly." Her sister said that she didn't suffer fools at all.

In 1994, a new arrangement allowed Canada's astronauts, heretofore payload specialists, to go through training to become mission specialists and thus career astronauts. Garneau, who had already flown, returned for an additional year of training as an astronaut candidate and joined the corps as a regular member. Bondar chose to return to Canada, where she could do her own medical research, rather than execute someone else's experiments, which is what mission specialists do. She left NASA as planned, but a decade late, after the single mission she had agreed to fly. She was the next to last female payload specialist and Canada's first woman in space.

The last woman to fly as a payload specialist, Chiaki Mukai, arrived as a long-term visitor to NASA from the Japanese Space Agency, which had negotiated an arrangement similar to Canada's. Chiaki's selection was an important signal that women had gained respect in Japan, and that inclusion in the astronaut corps was the pinnacle of achievement. By the end of the twentieth century those nations that could afford it wanted astronauts of their own, and that included Japan.

Before 1945 many Japanese men quarantined women in their homes or in small businesses that served only other women customers. Women played no role in the professional world and even spoke a separate "women's language." One of the consequences of Japan's defeat in 1945 was an effort to modernize both the economy and society. But it was only at the end of the twentieth century that Japan hastened to incorporate women in professional spheres. At the turn of the twenty-first century, one-fourth of its medical students were women, and many other women, including married women, worked in business and academia.

Following the Canadian pattern, the first Japanese astronaut scheduled to fly was a male, Mamoru Mohri. He went up in 1992 on an international mission carrying "Spacelab J," the "J" indicating the Japanese contribution. In 1989 the Japanese arranged for more missions, and as with the Canadians, the second Japanese astronaut, Chiaki Mukai, was a woman and a physician, but a cardiovascular surgeon rather than a neurologist. Mukai had been training since 1986, but she did not fly until 1992, when she flew on a fifteen-day microgravity mission. Unlike the others, she flew again in 1999, making her mark as the first Japanese woman in space and the first Japanese astronaut to fly twice. Mukai has remained a payload specialist and is still with NASA.

Hailed as Chiaki-san, the eldest of four children growing up in the 1950s, she lived in a small town that is now a Tokyo suburb, where her father taught in a junior high school and her mother ran her own shop. Always ambitious within the limits of Japanese society, she had, like Bondar, thought about becoming an astronaut. But Japan didn't have an astronaut program when she was growing up, and she did not waste time on an unachievable goal.

At the age of thirty-two, a physician on her way to becoming a specialist in cardiovascular surgery, she remembers reading an article that said the Japanese government was interested in space. But that was all it said. She continued her medical studies, completed her examinations as a cardiovascular surgeon in 1984, then she read another article announcing that there was a new Japanese space agency with ties to NASA—and that it would soon be looking for astronauts. She applied, one of 5,000 candidates, was accepted, and left for the United States. She, too, expected to fly in a year or two. Then came *Challenger,* and her mission was delayed for ten years.

During this uncertain decade she kept training in Houston, joined the faculty at Baylor University's Medical School and commuted to Japan, where she had left her husband, also a physician. She knows that this is not a conventional life for a Japanese woman, even for a woman doctor, but the public and the press seemed to like her unorthodox lifestyle.

She recalls that at first they kept asking, "How does it feel to be a female astronaut?" and she would reply, "Do you ask men how it feels to be a male astronaut?" She did not think there should be a difference, at least not in the way she felt, and she was startled and even insulted by the question. Then she realized that women had fewer career opportunities than men, and saw how crucial she was as a role model for Japanese girls.

She was in Houston during the *Challenger* accident and was still there three months later when there was a meltdown, killing and injuring thousands of people, at the nuclear power plant in Chernobyl in the Soviet Union. The two events are linked in her mind as evidence of technological, and thus human, vulnerability. But she never worries about space travel. As a doctor, she has seen that life is uncertain. "So many people think that the space program is very risky, but there are many more risky things that exist on this Earth."

Chiaki waited longer than the other astronauts with a pre-*Challenger* launch date to go into space, but in the long run she stayed in space longer and flew more often. Like the others who had been trapped by the *Challenger* disaster, she kept training, always ready and often disappointed. But when she was finally launched into orbit on STS-65, aboard the shuttle *Columbia* in 1994, she was in space for almost fifteen days, setting the record for a female astronaut on a single mission. STS-65 was the second microgravity laboratory mission and in its two weeks in orbit the crew completed eighty-two investigations of human physiology, space biology, radiation biology and medical experiments related to the cardiovascular system, the autonomic nervous system and bone and muscle metabolism.

During this mission as well as a second four years later on STS-95, Chiaki used herself as a subject in medical tests. She explained that as a doctor she had three responsibilities: to help her patients, to educate and to do research. By using herself as a test subject, she fulfilled the third. What she tested, among other things, was the impact of the hormone melatonin on her sleep cycle.†

She fulfilled her responsibility to educate by working with the same kinds of small plants that Japanese schoolchildren had grown as controls on Earth. She could see which plants grew "up" because of gravity, and which grew directed by some other, as yet unknown, process. Chiaki Mukai has remained a payload specialist and is scheduled for a third flight.

The barrier to women in the astronaut corps had dissolved surprisingly easily in 1978 when six women joined the first mixed class. Every class since then had included at least two women as mission specialists. In the mideighties there were still no women pilot astronauts, but women had entered the Navy and Air Force test pilot programs and a few were about to become eligible. By 1985, the number of female scientists and engineers in the United States was still very small, but female students had begun to enter medical schools in large numbers. Women were, in fact, embarking on a variety of professional careers in more than token numbers.

The first class after *Challenger,* which entered in June 1987, had been recruited and tested two years earlier, but like everything else in the astronaut corps, the final selection had been put on hold for a year. As was becoming a pattern, there were two women in this class. The first was Jan Davis, a mechanical engineer who had been designing rockets at the Marshal Space Flight Center in Huntsville. Like many incoming astronauts, she had been working at a NASA center before she applied to the astronaut corps.‡ The second was Mae Jemison. She had no connection at all with any space center, but like several of the other women in the corps, she was a doctor. She was also the first American woman of color to join the corps and the only one to fly as of 2003.

---

† The Japanese public, however, was more interested in her poetry than in her sleep. They were especially interested in the challenge she tossed to them: She had written the first seventeen syllables of a traditional thirty-one-syllable poem (a "Tanka"), which is split, like an English sonnet, the second half responding to the first. In the first half, she wrote "over and over in zero gravity." Five thousand four hundred people responded. One winner wrote: "How nice if I could bounce a ball of water." She was surprised to find many handicapped persons sending poems in which their bodies were whole again. Like Hughes-Fulford, she hopes that the work in space may someday help these people.

‡ Neil Armstrong was first hired as a propulsion engineer at Lewis (now the John Glenn Center) in Cleveland. When it became obvious that all he wanted to do was fly airplanes, he was transferred to the Dryden Flight Research Center located at Edwards Air Force Base in the Mohave Desert in California.

Mae Jemison had just begun work at the Cigna Health Plan in Los Angeles in 1986 when she returned home one day to find a thick letter from NASA. She had applied to the astronaut program a year earlier, heard nothing, and had gone on with her life. The letter said that NASA was still interested in her. Was she still interested, too? If she was, would she please fill in the enclosed form? She did, and several weeks later she got a letter inviting her to Houston. She was one of the final hundred of the 2,000 qualified applicants NASA would interview. Fifteen would become astronauts. In June she heard by telephone. It was George Abbey, then the head of astronaut selection in Houston. He wanted to know if she still wanted to be an astronaut. If she did, the job was hers.

Jemison said, "Yes," and like Judy Resnik ten years earlier, she quit her job in L.A., packed up her life, and, in her case, her cat, and moved to Houston. She did not know, of course, when orbital flights would resume. As soon as she arrived, she realized she was the only one in the first post-*Challenger* class of astronauts who was neither a member of the military nor a NASA employee. Instead of coming from universities or industry, like all of the women in the class of 1978, most new astronauts were coming from jobs at one of the NASA centers or through one of the armed services. This was a way of ensuring that the astronauts understood exactly how the space program worked, and it also ensured, to a degree, that future astronauts understood the degree of risk.

Even some of the physician-astronauts had begun their NASA careers as flight surgeons, the name given to physicians in the military. But Seddon and Fisher from the class of 1978 and Bondar, Mukai and now Jemison were civilian doctors. Jemison was like them in many ways. Like Seddon and Bondar, she had always wanted to be an astronaut. She had even studied Russian in high school to be prepared for what she correctly foresaw as joint Soviet–American programs. Like Seddon and Bondar she had noticed as a child that none of the astronauts was a woman, but she had also noted—and had taken very personally—the fact that none of the early astronauts, and none of the women after 1978, was black.

She was not, of course, the first to see this deficiency. By 1985 there were several African-American men in the astronaut corps. But until Jemison, there were no black women. Since the recruitment drive in 1976, NASA had been especially sensitive to gender and race and by 1985 was probably eager to recruit an African-American woman. However, NASA had not taken into account that an African-American candidate could have had a different experience growing up in the United States from her white colleagues in the seventies and eighties, a time when people of color had begun to seek out their heritage.

As a child with an older sister and brother, wise parents and an uncle who talked to her about Einstein's theory when she was barely six years old, she grew up wanting to become a scientist. She devoured science fiction, and with the exception of Madeline L'Engle's *A Wrinkle in Time,* she was dismayed to find women relegated to minor characters. "They were not scientists or adventurers, and when they were, they were not the heroes. To tell the truth, this pissed me off."

One of the first African-American students at a newly integrated Chicago high school, she dazzled her classmates as a cheerleader, dancer and president of the student council. She graduated at sixteen and chose Stanford over MIT, because of Stanford's football team and its financial package. She moved through college like any other academic prodigy, except that she was black and very aware that she had excelled in a recently all-white school. During the 1970s, against a backdrop of black empowerment, she investigated her personal heritage even as she worked hard at her major of bioengineering. Crucial to her education at Stanford was her discovery freshman year of African-American literature and the whole world of African studies. The publication of *Roots* by Alex Haley in 1976, the story of one black man's search for his ancestry through slavery back to the African homeland, must have validated her feelings and encouraged her to keep looking.

She never lost sight of her original goal, but when she graduated she had a double major in bioengineering, for which she had studied life science, space exploration and biomedical fluid mechanics, and in African and Afro-American Studies. At Stanford she had also performed in African and modern dance groups and had studied African politics and history. In her senior year she was elected president of the Black Student Union.

With these accomplishments she headed directly to Cornell Medical School in New York City, where she arrived in the fall of 1977. She recalls that she was "aged twenty and 5'9" tall weighing 136 pounds with two earrings in one ear and a feather in the other." She believed that she didn't look like either a New Yorker or a medical student, but she soon adjusted her "look," studied hard and finished on schedule, having spent the summer between her second and third years in Kenya. She hung out there with a group known as the "flying doctors," physicians and other Western health care professionals who traveled into remote parts of East Africa, providing surgery to isolated people.

After medical school she applied for a one-year rotating internship with no medical residency in mind because, as a bioengineer, she had never planned to practice medicine. Against the advice of the deans at Cornell, she spent the next year in Los Angeles at the County Hospital, and then applied to the Peace Corps. Accepted there, she took on a job where she was on call twenty-four hours a day, seven days a week in Liberia, across the African continent from Kenya. Eighteen months later she returned to Los Angeles, where she was contacted by NASA. It was 1986.

Jemison completed astronaut training in 1987 and in 1988 was in Florida to help with the launch of STS-26, the first mission after *Challenger,* when she heard on the radio that one of her heroes would soon be in town. Nichelle Nichols, the actress who had played Lieutenant Uhura on *Star Trek,* would be appearing at a Star Trek convention in Orlando during Labor Day weekend. Jemison set off to meet her.

The two African-American women, one a real astronaut, the other an actor who had played one, bonded immediately. A longtime trekkie, Jemison would begin all of her shifts when she flew on the shuttle *Endeavor* in September 1992 with a variant of Uhura's trademark greeting: "Huntsville, *Endeavor.* All hailing frequencies are open."

There were two teams of mission specialists on the *Endeavor,* as there had been on Bondar's mission. Each spelled the other so that the laboratory could be used around the clock. On Jemison's flight, one team was designated "red" and the other "blue." Jan Davis, Jemison's classmate, was also in this crew. And unknown to the astronaut office until right before the flight, Davis had just married Mark Lee, another crewmember, breaking NASA's unwritten rule against flying married couples. This rule was designed to make sure that no personal commitments inhibited the normal flow of crew relations. To limit contact, the lovers were assigned to different teams, and officially reported that they scarcely saw one another, except fleetingly, during changes of shift. Their presence, however, teased the imaginations of reporters on the space beat.

Both teams worked inside the spacelab module that filled the shuttle's entire cargo bay. Jemison's team performed forty-three experiments, including one with frog embryos, and another using herself and other members of the crew as test subjects to see how a biofeedback apparatus developed by Patricia Cowings at NASA Ames Research Center worked in orbit. For this experiment Jemison had to avoid any medication that would prevent Space Adaptation Syndrome. She was her own subject and found it hard to accept the decision of some of the other astronauts not to cooperate in the experiment. "A lot of astronauts don't want to [take part in experiments], but I thought [they should]. It was something they ought to get over." Jemison was not alone in her frustration. The lack of cooperation among astronauts in medical tests is a major reason for NASA's inability to build up a data base on the medical responses of humans in space. This issue would be papered over in the years following Jemison's mission, only to come to a head a decade later.

Astronaut testing was not Jemison's only frustration during her career with NASA. She and Bondar agree that the agency has a bias against single people. Twenty-five years earlier when NASA had chosen the Mercury Seven they had deliberately selected married men with children. They may have believed that marriage meant stability, although many of those marriages were not all that stable, or it may have been that NASA relished the image of an astronaut returning from space to a marital embrace and hugs and diapers. There have been a few bachelors and unmarried women in the corps, but no one in the first class of women complained of this bias. For whatever reason, Roberta Bondar felt slighted when "on the day we launched my family could not be in the special room [at the Kennedy Space Center] because I wasn't married." That was the special room in which Judy Resnik's family watched her fateful launch. After *Challenger* the protocol changed and no one at NASA seems able to explain when or how this happened, but Jemison found the same ban on nonspousal family.

Jemison also noted that "women [astronauts] put up with a lot more nonsense and tried very hard to be gender neutral, which was strange." Indeed, early feminists had fought the gender battle by arguing that there were no real differences between men and women and this point of view had become part of the catechism of many female astronauts. Even today, when asked, many of them insist that there is no greater difference between male and female astronauts than between two male astronauts or two female astronauts. Jemison looked at her male and female colleagues in disbelief, wondering how their own eyes could so fail them. She was not the only female astronaut who recognized essential differences between the sexes, but she had no idea how important this perception would become.

NASA was surprised when she resigned. They had selected a woman who could have come from Central Casting—brilliant, beautiful and charming. Then suddenly she was gone. But NASA shouldn't have been surprised. Unlike her colleagues, Jemison never identified herself primarily as an astronaut. Jemison left NASA after her flight—the only woman astronaut to quit after a single mission. She had worked hard and enjoyed the experience of weightlessness, of seeing the spectacular view of the Earth, but she was not "addicted to space." She wore other hats. Passionate as she was about space travel, she is passionate about space outside of the National Aeronautics and Space Administration.

Perhaps Jemison's vision of her future had always included a branching career path. Perhaps she was simply restless. After a mission, astronauts generally work at special assignments in Houston or at other NASA centers while waiting for another flight. Some have waited a decade, while others get one flight assignment after another. Sometimes the reasons are clear, but not always. Jemison chose to bypass the haze of uncertainty that shadows an astronaut's career because she had dreams beyond space exploration.

Unlike most of the other women astronauts, Jemison called herself a feminist. She also identified herself as a symbol for women of color. She recalled thinking, as she felt the engines shudder at liftoff, that she was not only the "first African-American woman in space, but the first woman of color in the world."[**] She looked at the fictional Lieutenant Uhura, the only woman of color on the Starship *Enterprise,* and saw a role model. Unlike Resnik, for instance, who wanted to be known *only* as an astronaut, Jemison wanted to be known as a female astronaut who was black.[††]

The *Challenger* disaster changed space travel. In the down years, as NASA reengineered the hardware, it reshaped the astronaut corps as well. Despite the devotion of Barbara Morgan, Christa McAuliffe's backup, a second teacher has not been sent into orbit. The role of payload specialist was left undetermined. When Morgan does fly in the twenty-first century, she will fly as a professional astronaut, a mission specialist with an educational specialty.

[**] This was two years before Chiaki Mukai would fly. There has not yet been an Asian-American female astronaut. The second African-American woman was scheduled for a mission in 2004.

[††] Pratiwi Sudarmono is still waiting in Indonesia. She was thirty-three years old in July 1985, when a visiting NASA team told her she would fly on the shuttle *Columbia,* STS–61H, scheduled to launch in June 1986. Her government had accepted an American offer to send an Indonesian astronaut to launch a second communication satellite, Palapa B–3, to supplement the work of the giant Palapa-B that NASA had rescued and relaunched in 1984. As with all partner nations, the Indonesians conducted their own selection process and, from a pool of 207 candidates, nominated four finalists, and finally, with NASA's cooperation, they selected Sudarmono. Like Chiaki Mukai, she had never fantasized about a career in space because it was not part of what seemed possible in Indonesia. But once the link with NASA was established, she wanted to fly. Like Bondar and Mukai, she trained in Houston in the spring of 1986. Then she returned home. She is still officially an astronaut candidate. Her flight was never cancelled. But after NASA launched Palapa-B robotically aboard a Delta rocket, she knew she was no longer needed. The Indonesian communication satellite was up, and Indonesia no longer held a major strategic role in American foreign policy. The narrow window of opportunity had closed for her.

Shuttles seemed safer after the redesign following *Challenger,* and the astronauts who flew them, aware of the problems implicit in such complicated machines, had faith in the reliability of the rebuilt fleet. No one expected a repeat of the O-ring episode, and gradually the perception of danger faded.

As the Cold War ended, NASA began preparations for sending its astronauts to the Russian space station *Mir.* Once again the Soviets had taken the lead in space exploration, in this instance in testing the capacity of two or three people to live and work together for extended periods of time in orbit. New records would be set as women and men from different cultures redefined their roles two hundred miles above the Earth.

8

NOT QUITE A HOTEL

IN THE EARLY MORNING OF February 20th, 1986, the Soviets launched a seventh space station, *Mir* (Russian for "peace" and also "world"), which settled into orbit 250 miles above the Earth. *Mir,* the successor to a whole series of space stations, replaced the successful *Salyut 7,* in which only two years earlier Svetlana Savitskaia, the second Soviet woman in space, had become the first woman to do a space walk. Besides setting a record, Savitskaia had set a precedent: Women would fly on Soviet space stations, and by extension on all space stations.

A month earlier and half a world away, *Challenger* had exploded, grounding America's shuttle fleet. It looked for a while as if the space race had flipped again, with the Soviets regaining the lead.

While the United States had spent the previous decade devising an ambitious double plan—a space station with reusable shuttles to ferry astronauts back and forth—budget pressures had obliged a compromise that meant postponing, but not forgetting, the wished-for space station, and just building shuttles. Even here the cost-cutting had continued, forcing redesigns to carry larger cargo in the misbegotten vision of selling laboratory space to industry and to other countries in the hope that the shuttles would be able to pay for themselves. It never worked as planned. Nevertheless, the elegant winged gliders were an engineering marvel in their ability to rise on rockets, orbit as a station, and return as a glider.

Meanwhile the Soviets, bowing to their own fiscal woes, took the opposite course. They opted to build new space stations and retain the reliable *Soyuz* to ferry its cosmonauts back and forth. *Mir* was not elegant like the shuttle, but it, too, boasted a new concept. Unlike its Soviet predecessors or NASA's Skylab, *Mir* was designed to grow. It began life in orbit with a single base block that had six docking ports. One port was immediately occupied by the *Soyuz* taxi craft, which remained docked as a lifeboat, and another port was soon used by an unmanned Progress cargo craft. The crew lived and worked in the base block but had access to the *Soyuz* for privacy. A year later, in July 1987, the first additional module docked, Kvant 1, with two telescopes and other scientific equipment. Two years after that, in 1989, a second module, Kvant 2, joined the station, with a new toilet and shower. Six months later the Kristall module arrived, a laboratory with equipment to process biological materials. Kristall also had a docking port that could, and would, be used by an American shuttle and two storable solar arrays—antennae that when extended collected energy from the sun. It had plenty of space compared to the little *Soyuz,* with a 39-foot-long chamber that provided both living and working space for the permanent crew—at a minimum two cosmonauts to run the station and a third visiting mission specialist.

Occasionally, in addition to the regular crew, a *Soyuz* would bring another two cosmonauts and perhaps a visitor. There was enough room, with the docked *Soyuz,* for six people to work and socialize. As an outpost in space, *Mir* served the Soviets, and then the Russians, as a place to continue research on the effects on cosmonauts of long-duration flight, of extended exposure to microgravity. It was also a kind of space hotel where the Soviets could host visitors and, like many hotels, sometimes turn a profit. It was a place in space that by its existence broadcast the generosity and technological superiority of the Soviet system. Also, not to mention, for it seldom was, *Mir* provided a perch for watching America's nuclear missiles.

The first visitors to *Mir* were guests from politically allied countries. But as the Soviet Union began to fall apart, the Soviets increasingly saw *Mir* as a source of cash. Looked at in the cold rationale of profit and loss, the Soviet space program—which by the late eighties largely meant the upkeep of the space station—was an unjustifiable luxury. As the Soviet Union took its last gasps, NPO Energia, the design bureau begun in the fifties, became a private company with public responsibilities and a very small public subsidy.

Eager for funds, the Soviets offered foreign countries, socialist and capitalist alike, their facilities to train cosmonauts in a process that took about eighteen months and cost roughly $20,000 a person. For another sum that would be negotiated separately but was in the area of $12 million, the Soviets would send the freshly trained visitor to *Mir.* In this climate of intense interest in space travel, the Soviets hoped to sell places on their space station. Meanwhile, as the station came to the end of its projected life span, problems intensified. For one thing, without money to develop new technologies or replace what had worn out, the cosmonauts had to work with a dearth of new parts and outdated instruments, especially computers. This would have been bad enough, but the program also suffered from some of the idiosyncrasies of Soviet, or perhaps Russian, culture.

Among these was its conflicted attitude toward women. The Soviets had always accepted, even depended on, women in the workplace. But acceptance did not necessarily mean respect, much less equality. Some Soviet men clung to chivalric gestures—welcoming Svetlana Savitskaia to the *Salyut* station with a bouquet of flowers—such as they were. Yet they also presented her with an apron, a symbol of her subservient, female status. Cosmonaut culture, perhaps an exaggeration of ordinary Soviet culture, was notoriously chauvinistic. At the turn of the twenty-first century, the disparagement of women cosmonauts had resulted in the demonization of Valentina Tereshkova. Forty years after her flight, every glitch—her failure to eat, the pill not swallowed—had grown in the retelling so that she is now said to have been disabled by nausea and gifted with an inability to do anything right. Feminine weakness, the Russians explain, led to incompetence, and this myth is still used by members of the Russian space program to explain and validate the miniscule number of female cosmonauts.

To be fair, the Soviet and, later, Russian reluctance to fly women can be seen as a reasonable response to the space limitations of the *Soyuz.* Not much bigger than the original space capsules, it carries at most three people seated very close together. There is little privacy. The *Mir* space station usually had a crew of two, but occasionally three, and often visitors. Soviet society was still puritanical in the second half of the twentieth century, at least when it came to men and women working together in such close quarters. The shuttle, in contrast, carries between five and eight astronauts in a relatively larger area. And since 1983, the shuttle crew has more often than not included one and sometimes two or even three female astronauts, the numbers changing the dynamics of a mixed crew.

Apart from unreconstructed sexism, the Soviet reluctance to fly women had less to do with sexual tension than with the competition for slots. There have always been fewer cosmonauts than astronauts, and fewer seats to fill. With only the *Soyuz* and *Salyut,* the cosmonauts, understandably, did not want to relinquish their few places to women. The same dynamic made the cosmonauts loathe to relinquish slots to foreign cosmonauts, which they were increasingly obliged to do. During its thirteen years in orbit, *Mir* flew sixteen Soviets and hosted sixty-four foreign visitors. Of the eighty people who spent long-term missions on *Mir,* only four were women, and only one was Russian, and she was not even first.

Mir had been in orbit for five years before Helen Sharman broke the gender barrier. A twenty-seven-year-old British food scientist at the Mars Candy Company, Sharman won the honor of being the first woman to visit any space station. It was a contest, with all the public relations events and photo ops that contests entail, and after all that razzle-dazzle she almost lost the prize of being first to a Japanese woman, who was also under contract to the Soviets.

This was still the Cold War, but money had already become a major factor, maybe *the* major factor, in the race to get a place in space. For even as British businessmen were negotiating with the Soviets to send their candidate to *Mir,* so was a Japanese broadcasting company that had selected two candidates: Toyohiro Akiyama, a chain-smoking senior news broadcaster, and Ryoko Kikuchi, a young camerawoman. As always with Soviet launches, one would fly, the other would serve as backup. It had looked doubtful that the broadcaster would make it, which led to the possibility that Sharman would not be the first woman. Akiyama had no detectable muscles and was in poor shape. But he managed to stop smoking, lose weight and he began to jog. Unhappily for him, his dramatic change of behavior did not prevent his susceptibility to Space Adaptation Syndrome. He vomited almost constantly during his week in orbit. A decade later, an American journalist in Russia heard the cosmonauts still marveling: "They had never seen anyone vomit that much." His condition led to pallid broadcasts from space, which was all the Japanese got for their 12-million dollar fee.

To the Soviets, however, the flight was a success—despite the fury the Soviet space officials had had to face from Soviet journalists who felt that one of them should have been the first to report from space. Interestingly, there has never been another journalist in space, not a Russian or an American, even though there was a lot of discussion about sending an American journalist before the decision was made to send a teacher instead.

Helen Sharman, next in line after Akiyama, had a much happier experience, as did her Soviet crew. When Sharman blasted off from Baikonur on May 18th, 1991, she was not so much the representative of a Soviet–capitalist partnership, as anticipated, but a guest of the Soviets. The British effort to get private investors failed and the Soviets, good sports in this instance, sent Sharman anyway to be paid in the intangible coin of goodwill.

Yet in that coin Sharman proved a wise, though somewhat improbable, choice. As she explains in her memoir, space travel had never been part of her dreams. In fact she had never given a thought to becoming an astronaut until the June evening in 1989 when, driving home with the car radio on, she heard an announcer say: "Astronaut wanted. No experience necessary." He explained that his company was looking for candidates, women and men, between twenty-one and forty, with formal science training, a proved flair for languages and good health. The winner would get to be the first Briton in space. The moment she turned up the volume to hear the rest of the announcement, her life changed forever. She later wrote, "I had no background in space research. I had never harboured ambitions to go into space, let alone be able to." She realized, however, that she fit the bill and managed to write down the phone number as she waited in traffic. From that day until the end of May 1991, she followed a bumpy road to fame.

Sharman describes herself as extremely ordinary, beginning with her birth, which had "no complications." She grew up near Sheffield, in central England, attended a local university, and had begun studying for a Ph.D. in chemistry at Birkbeck College in London when she heard that announcement. At that moment she had a good enough job, a good enough boyfriend and was happy, although she was restless enough to have been in the habit of scanning the employment opportunities in the newspaper. So she made the phone call, and dismissed the prospect until some months later when she was contacted, asked to fill out some lengthy forms and found herself in the hands of the Antequera Company, the sponsor of Project Juno.

Antequera had an impressive board of directors, including Gregory Pattie, a member of Parliament and a former space minister; he was responsible for coordinating marketing and making the final selection of the British astronaut and the British scientific experiments. The financial side of the flight was in the hands of a London insurance company, Jardine Glanville Interplanetary, which had once worked for the Soviet agency and had joined with the Moscow Narodny Bank to create Antequera, Ltd. The Moscow Narodny Bank, a commercial bank, had been founded in 1911 and nationalized by the Bolsheviks in 1918. With offices in London and Paris, it was subsidized by the Soviet state. A major link to capitalists abroad, it survived the Soviet era. In 1989, Antequera was supposed to raise $12 million to pay Energia for the flight to *Mir.* The price, though steep, would cover the expense of the *Soyuz* "taxi" and a week on the space station. Antequera was also expected to raise another $8 million to cover the expense of eighteen months of training for the two British finalists (like the Japanese and in true Soviet style, there were two British candidates, and the loser would be the backup) in the Gagarin Training Center in Star City.

The British government had made it clear from the start that they were not interested. Space travel would be a private venture, and Antequera was expected to raise money from British corporations that would then have the right to select the experiments that the British cosmonaut would perform.

The investors would harvest publicity, as well as scientific research, from tales of a British astronaut on *Mir.* In an age of unrestrained capitalism, Antequera was to prove that private capital could catapult the British into space.

Sharman's phone call to Antequera was one of 13,000 calls from all over Britain, from which Antequera winnowed out 120 candidates. Eventually she was directed to a Health Service office for physical testing, where she was hounded by the media. Sharman realized immediately that Antequera was using every step of the selection process, starting with the physical, as if it were a round of musical chairs; every time the music stopped and she was still seated, she was encircled by as many journalists as the company could arrange. She soon learned that the well-hyped selection was part of Antequera's effort to sell shares in Project Juno. The press had taken to calling her "the girl from Mars," a sobriquet she soon found tiresome. But by this time she wanted to fly very badly, so she put up with the "Mars girl" tag and cooperated in helping to sell Project Juno.

The whole enterprise seemed doable. It was simply a matter of selling shares to raise money to buy a seat on a rocketship. Sharman felt sure that the company would be successful, but not so sure about her own chances. It seemed obvious to almost everyone else in 1989 that the Soviets did not want to send a woman—perhaps out of fear that their own female cosmonauts, grounded in Star City, would complain. When the candidates shrank in number, the Soviets had to sublimate their prejudice and explain to the British that, for economic reasons, both finalists would have to be the same sex, perhaps to save on space suits or hotel rooms. When the number of British candidates fell to sixteen, of whom only two were female, it seemed unlikely that both would be chosen and that therefore neither finalist would be a woman. However, when Antequera further winnowed the list to four, Sharman's name was still there. In Moscow the remaining candidates went through medical examinations, and one was told to get his tonsils removed. Then there were three. The Soviets did not complain when it became probable that one of the pair would be a woman. Then they were down to two, and the public in both nations rejoiced. The finalists were Helen Sharman, the chocolate chemist, and Timothy Mace, an army pilot who was an aeronautical engineer and a member of the British National Parachuting Team.

Sharman and Mace had been training in Moscow for over a year when they learned that their contact in London at Antequera had resigned, that Antequera had not been able to raise enough money and, as a consequence, that the firm had fallen apart, taking with it Project Juno. What should they do under these circumstances? They did not pack up and leave but continued going to their classes in Star City. No one there seems to have discouraged them. They were already an investment, and perhaps a propaganda opportunity. Before long the Soviets arranged for the Narodny Bank to take over the program. The bank agreed to send a British astronaut, but abandoned the British scientific program. It is difficult to understand this generosity on the part of the Soviets. They may have wanted to salvage something from their efforts, like goodwill. Or maybe they just liked Sharman and Mace and wanted to send one of them into space.

Anatoli Artsebarksy, the Soviet cosmonaut who would accompany whomever was chosen, knew whom he wanted to fly with. He had counted on Tim Mace, the famous British pilot, being selected and had dismissed the possibility of a female partner, declaring: "It is not a woman's business to fly into space. More work can be done by a man." In an interview with the British monthly *Spaceflight* eleven years later, he explained what he had really meant. During their eighteen months of training, he had worked with Sharman and Mace at Star City. True, Mace had been his odds-on favorite because they had so much in common. He, like Mace, had been a military pilot and he believed that Mace, as a champion pilot, had better qualifications than Sharman. He both denied the disparaging remarks he had made a decade earlier, and defended them at the same time. He had felt more confident flying with another pilot, a person who had invested time and energy and had sacrificed to fill his dream of space travel. If it had been up to him, he would likely have chosen Mace.

But he did not do the choosing. Those who did, he believes, had a political agenda. They "decided that the publicity would be better if they sent a lady rather than a man." Whatever the criteria, Sharman got the nod and things moved quickly . She and Mace, along with her two Russian crewmates, Artsebarksy and his flight engineer Sergei Krikalev, and the deputy director in charge of crew training, Alexei Leonov, the first man to walk in space, flew from Moscow to Baikonur. They all stayed at the "cosmonauts hotel," where crews traditionally spend the night before a flight. There Sharman learned cosmonaut traditions, all echoes of Gagarin's 1961 flight. Like Gagarin she signed her name on the back of the door to her suite. She then rode with the cosmonauts who had come for the launch to a ceremony near the launch site. Before they boarded, Artsebarksy showed Sharman where to hide personal items inside her jumpsuit.

On the bus, Leonov gave Sharman a frilly pink chiffon jumpsuit. He told her it was to wear on the station, and she hid it inside her suit. Then there were good-byes and without any fanfare she found herself zipped into the pressure suit for takeoff, then strapped into her special form-fitted seat in the *Soyuz* alongside her crewmates.

A Soviet launch, she discovered, differed from the NASA launches she had watched broadcast from Florida. The Soviets had no dramatic countdown and liftoff because, in the Soviet Union, launches had not usually been public events.

But this did not lessen the excitement for her, and for her family who had come to Kazakhstan, and for the British public following her flight on television. After eighteen months of training, Sharman was ready. Did she think about Christa McAuliffe, an ordinary woman like herself chosen in a similar public contest? Not really, Sharman explains, because Christa did not have a Christa to think of. Besides, she felt safe in the *Soyuz.* It had never lost a passenger on launch.

Once in orbit, they unstrapped and drifted around, stretching their bodies in the small space. Forty-eight hours later the *Soyuz* docked with *Mir,* and the Soviet crew already there, Viktor Afanasyev and Musa Manarov, honored Sharman by inviting her to enter the station first. She floated in and was greeted with hugs and a scaled-down version of the traditional Russian bread and salt welcome to a new home. By the time she arrived, the station had been in constant use since 1986, and despite a great deal of Velcro and elastic bands holding things in place, her first impression was one of immense clutter.

On board, her colleagues offered her a bedroom to herself, which they did not offer visiting men. Occasionally, however, she changed her clothes right there in base block; Artsebarksy reported being uncomfortable that "she can change clothes in our presence." Then there was the frilly pink outfit. When she wore it to surprise her commander, he surprised her by declaring that it was not a surprise at all, that in fact he himself had had it made to order for her. But what did the outfit mean, if anything? It can be seen as a gesture to her femininity, or it can be seen as an admonition that a woman should be feminine and not dress like one of the boys. Perhaps it was simply a gift, Russian-style.

Sharman claims that she did not notice any hostility toward her as a woman, and it is likely that her trainers were all practicing good behavior. Yet only the year before Alexander Alexandrov, head of Energia's civilian cosmonaut training program, expressed the quasi-official attitude, in TASS, that females had no place in the cosmonaut program. Yes, he acknowledged, the Soviets had sent Valentina Tereshkova into orbit thirty years earlier, but they now realized that things had changed. A task like a spacewalk (such as the one that Svetlana Savitskaia had brought off so well) "is demanding work, even for men." He said that he did not rule out sending women experts on short flights in the future. But he continued, changing the argument from brawn to behavior, "the long-term stay of men and women in orbit together brings about moral and ethical problems."

Short flights, he had to concede, had been no problem. In the light of Savitskaia's unblemished scientific and engineering success and her upright deportment in 1984, and, he noted, the squeaky-clean behavior of the dozen NASA women—who had always been a minority of one or two in a crew of five or more men—short-term, heterosexual flights couldn't be faulted. In fact, he suggested, in the short term, the presence of a woman had a morally uplifting affect on her male companions, who, with a woman present, shaved regularly, spruced up their clothing and ate with whatever decorum is possible when liquids do not pour and solid matter can drift away.

In his conversation with *Spaceflight* a decade later, Artsebarsky concedes that, while he had preferred to fly with Mace, Sharman's scientific work was perfect: He didn't "remember her making one single mistake." However, at the time he complained in a widely broadcast press conference conducted from orbit that "It's hard work, not a woman's work. It is impossible to understand if she is in a good or bad mood." He dismissed her inscrutability as a matter of gender, ignoring the language and cultural divide between them.

Like all cosmonauts, but not astronauts, Sharman was ordered to carry out medical experiments on herself, including monitoring her heart with electrodes for twenty-four hours and taking blood samples by pricking her fingertip twelve times a day. She was doing this, she explained, because most cosmonauts had spent months in space, leaving the medical team with sparse data from short-term visitors. Her tests helped fill this data gap. Another medical gap she helped fill was testing the air quality on *Mir.* Dust and food particles floated everywhere and, consequently, the cosmonauts sneezed often and fitfully. Sharman gathered filter papers with samples that she brought back with her for examination on Earth.

She did do a few experiments for those British investors who had gambled on Project Juno. Suttons, a seed company, had provided her with 125,000 seeds, half of which she left behind and the rest brought with her into space. She later distributed the seeds to schoolchildren, giving each a set of seeds from both groups, the seeds left behind and those that had orbited. The children could then watch them grow and compare the plants. Besides conducting laboratory work for the Russians, she also grew protein crystals, which X-ray crystallographers analyzed on her return. Protein crystals form more rapidly and with fewer impurities in zero gravity, tantalizing manufacturers with visions of factories in space, and more practically helping researchers seeking diagnostic uses.

The world of science and technology was not enormously enriched by her week's work, but the British public, and perhaps the British national ego, was. Sharman's good nature and flawless performance compensated for her awkward position. Although she had flipped from being a paying visitor to an expensive guest, she won the affection and admiration of her Soviet colleagues and the British public. She returned a celebrity, visiting schools and speaking to public and private organizations, and in 1993 she published her co-authored memoir. She had "seized the moment" (the title of the book), but it was only an episode in her life, and she closed that chapter quickly, no longer a cosmonaut and public figure.

The demise of Project Juno and its parent company, Antequera, provided a lesson about the limits of private enterprise in space travel, at least the kind of private enterprise that did not invest in its own launch pads and spacecraft. The choice of Sharman suggests that by 1991 it was politically advantageous to launch a woman. The British clearly understood that and so did the Soviets, who helped make the choice. But the latter saw it as useful in international relations. They had already sent up two female cosmonauts of their own and were not, apparently, pressed to make that gesture more than once a decade.

Sharman did not want to be called a "woman astronaut," and she never mentions, and perhaps never noticed, the chauvinism endemic in the cosmonaut corps. If she had understood Artsebarksy's remarks, she did not think they applied to her. On *Mir* no one greeted her as Svetlana Savitskaia had been greeted in 1982. There were no flowers. On the other hand, there was no apron either. As for the pink chiffon, whatever it meant, she went along with it. The crew obviously liked her, but they had no reason not to. She did not threaten their places on future flights, and since the unstated purpose of the mission was goodwill, her Soviet colleagues did their job and welcomed her. They succeeded in helping Sharman enjoy her every moment in space. In appreciation she declared her Soviet companions "Quite simply, the closest and most important friends I had ever had."

Sharman was both the first woman and the last visitor to *Mir* in the Soviet era.

Shortly after her return, the Soviet Union disintegrated. From its ruins a smaller Russia emerged. Fiscally challenged, it was eager to regain preeminence on the world stage.

Mir remained in orbit as Russia adjusted to a new kind of state. The commercial adventures with Japanese and British partners had had mixed results under the Soviet Union; now as Russians, they expanded the use of the space station as a source of income. All of these changes, understandably, had left the semiprivate agency Energia in an awkward position. It needed money, and perhaps a new partner.

In 1991, before the Soviet Union disappeared, the advantages of cooperation between the only two human space programs seemed obvious. That year President George H. W. Bush and Premier Mikhail Gorbachev signed an agreement calling for a U.S. astronaut to fly on a *Soyuz,* and a Russian cosmonaut to fly on a shuttle. They also discussed the possibility of the United States building a module that could dock with *Mir* on a future flight.

A year later, in 1992, NASA's new administrator, Daniel Goldin, arranged an elaborate agreement between the two nations that would evolve into the Shuttle–*Mir* program. Phase 1 of Shuttle-*Mir* involved sending seven astronauts to visit *Mir* for about four months each. In order to prepare for the visit, the Americans would have to live in Russia for about a year to learn enough Russian to train with the Russian crew. They would then be able to visit the space station and learn, firsthand, about the limitations and strengths of the Russian program. They would discover what it was like to spend a long time in space with people from a different culture who spoke a different language. Not incidentally, they would work out the problems of working with Russians. In exchange NASA would pay the Russian Space Agency and NPO Energia enough money to keep *Mir* in orbit.

NASA did not know much about the physical condition of *Mir,* except that Russia had expected it to remain in orbit for five years and it was already long past that expiration date. *Mir* would keep orbiting until 1999, more than twice its estimated lifetime. As NASA and NPO Energia worked out the details of the Shuttle-*Mir* exchange, the Russians decided, a decade after Savitskaia, to orbit another Russian woman. In the three years since Helen Sharman's mission, the Russians had sent eighty-four male cosmonauts to *Mir.* Energia may have been operating under a Russian flag, but it had inherited the personnel and the tactics of its Soviet progenitors. A new record was about to be set.

By this time the Russians had stopped thinking in terms of an all-female crew. Yet when they announced that a third Russian woman, Elena Kondakova, would fly as the equivalent of an American mission specialist, it was clear that it would be no ordinary mission. Like Tereshkova and Savitskaia, Kondakova would set a record. The agency was no longer concerned about the morality of leaving a woman alone with a man who was not her husband in the small confines of *Mir.* Kondakova, an engineer, would remain in space longer than any other woman, leaving Kazakhstan on October 4, 1994, and orbiting the Earth for 169 days before returning on March 22, 1995.

Elena Kondakova, born in the year of *Sputnik,* was thirty-seven years old when she visited *Mir* in 1994. She did not have an influential father like Savitskaia. Her special connection, which she said played no part in her selection, was her marriage to Valerii Ryumin, a cosmonaut who had flown as flight director on the *Salyut 6* space station and had set his own record orbiting the Earth for more than 300 days. Kondakova is wise in the sexual politics of her generation. She knows how Russian men think about women, and she's skilled at getting concessions and presenting herself, and the Russian space program, to the world. Tall, with an athlete's build, she fitted easily into the Russian space suit. When she went to *Mir* she left her eight-year-old daughter in the care of her formidable husband—a twice-named Hero of the Soviet Union and since 1992 director of the Russian side of the Shuttle–*Mir* program. He had married Kondakova, his second wife, in 1985. In Star City and before the world's press he had a reputation for a volcanic temper and a good sense of humor, the latter quality dominating, it seems, in both his marriage and his relations with NASA.

Kondakova had not grown up wistfully dreaming of traveling in space. For her, deciding to become a cosmonaut had been akin to deciding to enter the family business when there was a place for her. She had grown up near Kaliningrad, the home of Korolev's design bureau, which became the private NPO Energia with the end of communism. Both of her parents worked there. She was nine years younger than Svetlana Savitskaia, who as the daughter of a marshal in the Air Force was a member of what passed for an aristocracy in the Soviet world. Kondakova's parents were Moscow-centered, middle-level engineers. Her father headed one of Energia's labs, and her mother worked there as an accountant. Space travel was the world she had always known. If other aspiring cosmonauts grew up inspired by science fiction, Kondakova bristles at the suggestion that fiction could ever have influenced her career. "I lived in a real city where real people worked, real cosmonauts returned after their flights. I saw around me a real thing. We were very proud that it was in our city that spacecraft were built and cosmonauts trained."

But when she was growing up, most of these "real cosmonauts" were men. She had not considered space travel as a career when she followed her older brother to the Bauman Technical School in Moscow, where she trained and worked as an engineer. Her choice of Bauman pleased her parents, she explains, because "they knew that I was mathematically and technically inclined."

"You're just like the rest of the country—prejudiced against a woman astronaut!"

Valentina Tereshkova inspects the Apollo Command Module on exhibit at the Johnson Space Center with Apollo Astronaut Alan Bean in 1977. (Courtesy of NASA.)

Valentina Ponomareva at the Gagarin Cosmonaut Training Center (Star City) in Russia in 1999. (Bettyann Holtzmann Kevles.)

Svetlana Savitskaia. (Smithsonian Institution, National Air and Space Museum Photo No. 83-11255. Courtesy of The National Space Society.)

Sally K. Ride aboard Space Shuttle *Challenger*, 16-24 June 1983. This photograph triggered angry letters to NASA criticizing the astronaut's short shorts that exposed her bare thighs to television cameras—and the world.
(Courtesy of NASA.)

Anna L. Fisher aboard Space Shuttle *Discovery*, 11 November 1984.
(Courtesy of NASA.)

Reprinted with special permission of King Features syndicate.

Kathyrn D. Sullivan onboard Space Shuttle *Challenger*, 14 October 1984, using binoculars through forward cabin windows.
(Courtesy of NASA.)

Mary L. Cleave performing materials science test onboard Space Shuttle *Atlantis*, 1-8 May 1989. (Courtesy of NASA.)

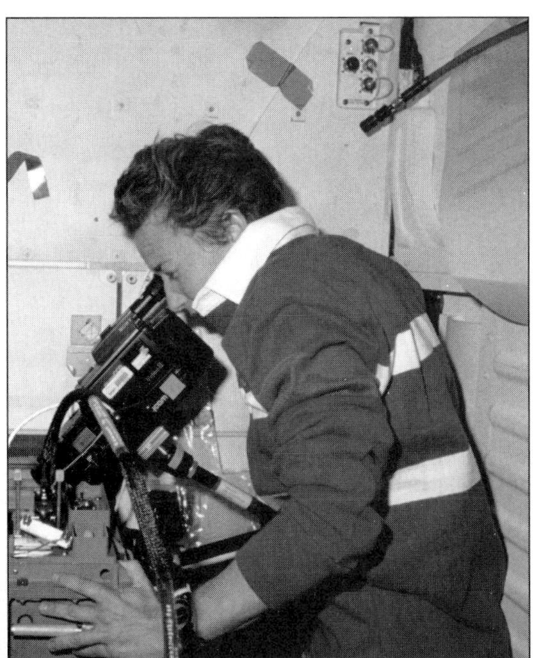

Bonnie Dunbar in a cosmonaut space suit in the Training Simulator Facility at the Gagarin Cosmonaut Training Center (Star City), March 1995. (Courtesy of NASA.)

"Damn this light gravity—it effects everything..!"

Rhea Seddon at her office at Vanderbilt University School of Medicine 1999. (Bettyann Holtzmann Kevles.)

Judith A. Resnik peeking at mid-deck activity onboard Space Shuttle *Discovery*, 6 September 1984. (Courtesy of NASA.)

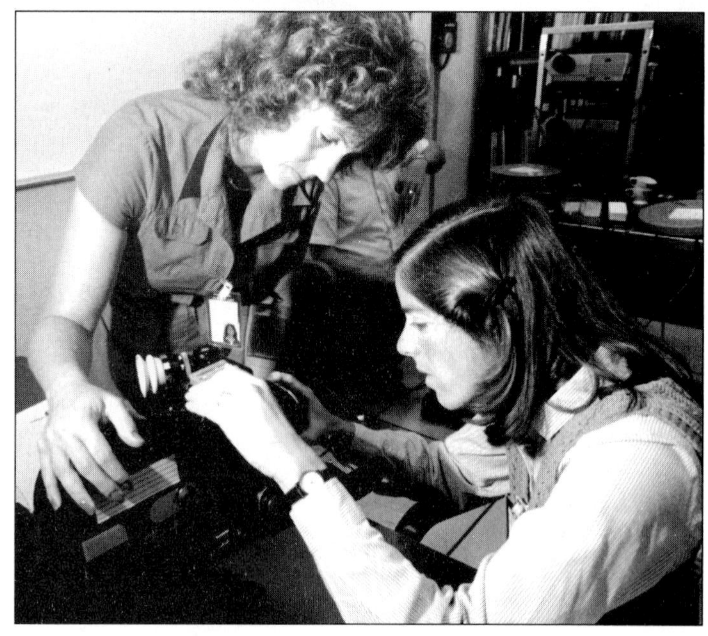

Christa McAuliffe and her backup Barbara R. Morgan practice using a motion picture camera while training for STS flight 51-L, *Challenger,* 18 September 1985, (Courtesy of NASA.)

Shannon W. Lucid in her Office as Chief Scientist at NASA Headquarters in Washington, May 2002.
(Bettyann Holtzmann Kevles.)

Claudie Heigneré in her office in Korolev, Russia, 8 August 1999.
(Bettyann Holtzmann Kevles.)

Elena V. Kondakova working in the Spacehab glovebox onboard Space Shuttle *Atlantis*, 15-24 May, 1997.
(Courtesy of NASA.)

Chiaki Mukai enters the Microgravity Laboratory science module via the tunnel connecting it to the cabin of the Space Shuttle *Columbia*, July 8-23 1994. (Courtesy of NASA.)

Mae Jemison in Spacelab-J in the cargo bay of Space Shuttle *Endeavour,* July 25, 1992, before lift-off in September. (Courtesy of NASA.)

Millie Hughes-Fulford in her office at the Veteran's Administration Hospital in San Francisco, April 1999. (Bettyann Holtzmann Kevles.)

Janice E. Voss floating through the tunnel into the Spacehab module onboard Space Shuttle *Endeavor*, 21 June-21 July 1, 1993. (Courtesy of NASA.)

Pamela A. Melroy holding camera equipment as she floats into the Zvezda Service module on the International Space Station, 1-18 October 2002. (Courtesy of NASA.)

(left) Susan J. Helms on board the Space Shuttle *Discovery*, 21 August 2001, dressed in full-pressure Launch and Reentry suit as she prepares to return after five months on the International Space Station. ((Courtesy of NASA.)

(below) Eileen Collins, commander, participating in a simulation of emergency egress from a Space Shuttle in the deep Neutral Buoyancy Laboratory pool at the Johnson Space Center's Sonny Carter Training Facility, June 14, 1998. (Courtesy of NASA.)

Catherine G. Coleman using the bicycle ergometer on the flight deck of Space Shuttle *Columbia*, September 5 - October 20, 1995. (Courtesy of NASA.)

Ellen S. Baker near the docking port of the *Mir* Space Station, posing with Cosmonaut Vladimir N. Dezhurov. June 21- July 5, 1995 (Courtesy of NASA.)

Kalpana Chawla (foreground) and Laurel B. Clark work in the Spacehab Research Double Module onboard the Space Shuttle *Columbia*, 17 January 2003. This picture was sent electronically from space at the start of the mission that ended two weeks later in the destruction of the Shuttle and the deaths of the entire crew. (Courtesy of NASA.)

What she did not do was follow the path of her female predecessors. She did not learn to jump with a parachute like Tereshkova, or pilot a plane like Savitskaia. She preferred hiking and fishing for sport, and reading and going to the theater for entertainment. She did not seek thrills or venture far from home. So with her new degree in engineering in 1980, she found a job at Energia and in 1985 married Ryumin, one of the legends of her world. While an engineer at Energia, she learned that for the first time both women and men would be considered for the next class of cosmonauts. She applied with several of her friends in the company, the only woman among the four new cosmonaut candidates. But, she recalls, she had never worried about the tests. What concerned her was Ryumin. As deputy chief designer at NPO Energia he had once signed an order forbidding women cosmonauts assignment on long-duration missions.

"My husband vehemently opposed this idea. He had already been in space three times, [three flights total] a year." He finally agreed to let her apply, she explains, because "he was convinced that I would not pass the medical tests." When she did pass them, he told her "I never doubted that I had a healthy wife." He was not pleased with her success.

Nor were her parents encouraging about her decision to become a cosmonaut. "My father worked in this area and dealt with actual tests. He knew how really dangerous it was." And he did not want his only daughter in danger.

Ignoring the total lack of familial encouragement, she entered the cosmonaut corps anyway, began training, paused to have a daughter and returned to work. Whether or not she acknowledges it, she enjoyed an umbrella of protection. Ryumin may not have approved of her new position—he protested loudly and often—but eventually he appeared to accept her decision. But because of who he was, no one in Star City would have dared treat her with anything short of respect.

Ryumin did, however, have to make some concessions for her career. Life in Russia, as Kondakova pointed out, is a life without baby-sitters. The fabled "babushka" has a day job now, and Russian teenagers study, they do not baby-sit. As the only woman in the corps, she was allowed to limit her working hours in order to drive her daughter to and from school in Moscow, where she and Ryumin lived in greater than average comfort. Cosmonauts, at least those with prestigious positions, always enjoyed special privileges, including apartments in Star City or, in the Kondakova–Ryumin case, a town house in a gated cosmonaut community in Moscow.

When Kondakova's mission was announced, she and Ryumin played out the equivalent of a sitcom before the world's press. Unlike the many astronauts who are married to astronauts but seldom mention their spouses, Kondakova and Ryumin seemed to revel in explaining their marital discord. They provided separate but complimentary interviews to amuse the American press. Ryumin maintained that he would have preferred a wife who worked only a few hours a day, took care of their child and always had dinner waiting for him. She agreed with him. She insisted that she shared this image of the way a marriage ought to work, but in their case, unhappily, she simply could not oblige.

When she left home, Energia expected her to spend four months on *Mir* to set an endurance record for a woman in space. Energia then reassigned Ryumin to his home to take care of their daughter. In the topsy-turvy world of the new Russia, this was apparently easier than finding a housekeeper, and perhaps it was what Ryumin really wanted to do.

While training, Kondakova got to know some of the American astronauts and their families who had just come to Star City as part of the new cooperative program. Kondakova had followed the American women's movement in the news a decade earlier and laughs at the suggestion that she had not known about it. "We did not live behind the kind of iron curtain you imagine. Now, after twenty years, I think that maybe we could have organized something [for ourselves here] but it would have been illegal." She thinks about it some more. "We would have liked to participate in this kind of thing, but in real life, it is clear that the [women's movement] is the product of the American mentality." What she means is that Russian and American women have not only different ideas of the way men and women get along, but different histories and, above all, different strategies. The Russians like to think of themselves as more French than Anglo-Saxon, more flirtatious than Americans in social situations, while being as effective as men in a factory or a space station.

She knew about the proposed all-woman flight in the mideighties and knew "that the men [the cosmonauts] were all vehemently opposed to it." She recalls that Savitskaia, the only woman with two flights under her belt and thus the only woman qualified, was supposed to have been the commander. They had already started training "but she walked out to give birth. She decided that it was better to have a baby than to have a third spaceflight." It is hard to tell what Kondakova thinks about Savitskaia's decision, and harder to know whether there was really any chance for that long-promised all-female flight to happen. By the nineties, with the shuttle flying again, often with a woman or two in the crew, the idea of an all-female crew seemed strangely old-fashioned.

It is not hard to sense Kondakova's opinion of her predecessor. She blames Savitskaia for the resentment she stirred among the men in the cosmonaut corps and describes her as having "a typical American" attitude, which probably refers to her determination to do everything a man could do on the station, an attitude that led to problems and misunderstanding on *Mir.* Kondakova explains: "For example, Savitskaia is a very good pilot and a very good engineer. No problems with technical expertise here. But in work relationships with men, it's a case of diamond cut [against] diamond. She wanted to take more responsibility than men may have been willing to give her." It came down to a question of strong personalities and a lack of diplomacy. She speaks from experience when she says, "One of the most important things in space is to be able to find a compromise."

It is easy to conclude that Russian women are simply a generation behind Americans in their attitude toward having careers. Kondakova disagrees. She believes that Russian women feel a greater responsibility to husband and home than do Americans. She suggests that is because she, and by extension other Russians, believe that "men and women are completely different." There are physical differences and also psychological differences. She says that she fully supports her husband's position that a wife must devote most of her time to her home and family and should work only three days a week and only until lunchtime: In this ideal world she would be home every day preparing a family meal.

When Ryumin explained this view to American women, Kondakova said, they were very upset. Defending him she said, "I do not think this is chauvinism. Such is life. When the husband comes home he wants to be met by his wife, he wants to see his children already fed, so that he won't have to cook his dinner himself after he comes home from work. I think this is normal." In her case, it is different. Her husband ends up cooking for her because she often comes home later. Then listening to what she has said she admits, "Maybe I think he is right because I feel guilty."

She enjoys being an exception and so defends the rule. In a sense she has it both ways. She stands up for her husband and the reputation of all Russian men by recalling that the American women she knows enjoy socializing with Russian men because, she suggests, they are more gallant, they place women on a pedestal and treat them "like ladies." She finds American women unreasonable: "In the U.S. women demand that they be treated as equal partners." She clarifies her point: "I am not speaking of the workplace. I speak of the time after work."

Yet on *Mir* there was no "after work," and Kondakova had to be all business all the time, which may have been a strain for her. During Sharman's week in orbit, the crew was on very good behavior because of all the international attention. Kondakova's crew also behaved well during her 185 days in orbit. Of course she was, and is, the wife of an important man.

Part of her work was simply living as normally as possible in zero gravity, and there were no special accommodations made for her. She did not, as do many American women, take birth control pills to eliminate menstruation and reports that her period "was easier and more regular than on Earth." She had left a child at home and did not worry about the impact of exposure to radiation on a second pregnancy, should she have wanted one. She refers to the cosmonaut Sergei Krikalev, who "fathered his child the first night after his return from space. Our child was also born after my husband had spent a year in space. There are many children in our group that were conceived after spaceflights." She did not consider that her own reproductive system might have reacted differently, that radiation might damage her eggs while leaving sperm, which are constantly being manufactured in the male, relatively unaffected.

She was acutely aware that accidents happen. "I was scared," she admitted. "We are human. There are things to fear when flying a plane or making a parachute jump. When a rocket starts, any normal person recalls the fate of *Challenger*." And returning to Earth reminded every cosmonaut of colleagues who had died when a parachute failed to open or, in another case, when the landing capsule depressurized, killing the three crewmembers aboard.

Elena Kondakova was still in orbit on February 3, 1995, when STS–63, the space shuttle *Discovery*, blasted off from Florida to meet the Russian space station *Mir*. It would be the first Russian–American space rendezvous since 1975 when an Apollo capsule docked with a *Soyuz* in a gesture of extraterrestrial friendship. Now *Discovery*, commanded by Jim Wetherbee with Eileen Collins, as pilot, at his side, was en route to a second international meeting. The plan called for the shuttle to circle the space station at a distance of about forty feet.

However, on February 7, four days into their flight, Wetherbee learned that *Discovery* was leaking nitrogen tetroxide. This was not unusual and would not have been a problem in a normal orbit, but it was not the kind of gift to bring along on a visit. NASA could not risk exposing *Mir* to a cloud of caustic fuel.

Collins worked desperately to stop the leak, and just when it looked as if they might have to postpone the rendezvous, she succeeded. Then, as planned, the orbiter hovered about thirty feet below *Mir*'s Kristall module's docking port and Wetherbee, the commander, addressed the Russians: "As we are bringing our spaceships together, we are bringing our nations together."

He read his carefully prepared text and then he recalls, "All of a sudden, we looked up, and we could see their faces. The interesting thing to me—and

I'll always remember this for the rest of my life—is that the first time we met them was *not* on Earth. And I saw Elena Kondakova waving to us, and she held up the little doll . . . a little cosmonaut or astronaut doll. And she held it up to the window. At thirty feet you could see their eyes and their smiles and they were waving and we were waving."*

The Cold War had continued into the early nineties, but as it waned NASA managed to interest international collaborators in the peaceful uses of space. By the end of the twentieth century this meant building a new international space station. The first part of this plan involved sending astronauts to work with cosmonauts on *Mir* to learn the problems and the Russian solutions to living in orbit for extended periods of time. The second part was constructing the International Space Station, and between 1998 and 2000 the largest number of shuttle missions were devoted to this task. Four times larger than *Mir,* the Station, if and when completed, will measure 356 feet across and 290 feet long with almost an acre of solar panels to provide electrical power. It is already one of the greatest engineering feats in history.

---

* Wetherbee was caught up in the memory of that meeting. As he recalled: "And the second time I met [Kondakova] was on Earth . . . at the Cape in Florida giving a briefing to the guests who were there to view a launch. . . . And I looked down in the audience and I saw Elena Kondakova—for the first time on the planet. So I told the crowd. In the middle of my speech I suddenly realized this. I tried not to sound too excited. And I just added, 'And by the way, the first time I ever met Elena was not on the planet. The first time that I'll meet her on the planet, if she will indulge me, will be as soon as I finish this speech because she's sitting in the audience!'"

9

CHATTERBOX IN ORBIT

IN PREPARATION FOR the International Space Station, the United States entered an extraordinary partnership with Russia in 1994. The United States would pay $400 million in "rent" so that the Russians could extend *Mir's* life. In return, the shuttle would bring seven American astronauts to the Russian space station to let them experience a long-duration mission. This arrangement, Shuttle–*Mir,* was labeled Phase 1 of the plan for an International Space Station, and a special docking structure was made for the shuttle on *Mir.*

The seven astronauts who flew learned about life on a space station and they learned especially about working with people who spoke a different language and came from a different culture. NASA had never paid much attention to these things—or even to the psychological makeup of its shuttle crews—because all astronauts spoke English and shuttle missions never lasted more than two weeks. The new International Space Station promised long tours of duty and experience on *Mir* would allow NASA to see how crew members managed in this very different environment.

In March 1995, Norman Thagard, the first of the seven astronauts, flew in a *Soyuz* from Baikonur to the space station. Three months later the shuttle *Atlantis* picked him up and brought him home to the United States. His visit was a triumph for NASA, but it had not been easy to arrange.

The American agency had needed astronauts willing to train in Russia for about a year and a half, and there were not many volunteers. Most of the 120 members of the astronaut corps did not want to uproot their families or family businesses. Besides, Russia did not have a good reputation in Houston. Earlier visitors had described a quasi-third-world country lacking conveniences and creature comforts. Beyond the matter of physical hardship, Russian is a difficult language to learn and few people in Russia spoke English. The list of willing astronauts was short, but it did include several women.

The first American woman, Bonnie Dunbar, arrived in Russia in 1994 as the backup to Norman Thagard, who would inaugurate the new Shuttle–*Mir* program. Dunbar, from the class of 1980, had already studied some Russian and as Thagard's backup she may have assumed she would be assigned to the next Shuttle–*Mir* mission. That was the way it usually worked in Russia.

Dunbar did not mind the spartan conditions in Star City, as some of the other Americans did. She had lived without indoor plumbing on her family's Washington homestead until she was a teenager, she told a NASA historian, and she found Russia interesting. She had been in the Soviet Union once before, in 1991—at a microgravity conference in Tashkent, Uzbekistan—and afterward she had accepted an invitation to tour Star City. She had liked what she saw.

Now, she explained, NASA needed volunteers, and as the daughter of a marine she felt it was her duty to go where she was needed; so she went. She had not anticipated the difference between being a guest for a few days and actually living with the cosmonauts. Before she left home she had not thought of herself as especially political, and although she had suffered some gender bias as an undergraduate at the University of Washington in the seventies, she had never identified herself as a feminist. But when she met the same kind of condescension from men at the Gagarin Training Center, she discovered that the old hurt had not gone away and she could not—would not—laugh it off.

Nor would she keep her observations about the paucity of women cosmonauts to herself. She kept seeing Russian women being slighted, and experienced instances of bias firsthand. The first affront was finding herself barred from the cosmonaut's gym because there was no women's dressing room. Next, in the classroom where she sat every day for training, she listened in fury to a Russian doctor who told her pointedly that if she actually flew on *Mir,* which he did not think would happen, she would have to have her own toilet. Because, he said, with no scientific data to prove it, "female urine had a different chemistry than male urine, and the *Mir*'s toilet could not process it."*

She despised the way Russian men seemed dismissive and condescending toward women in general, but she was especially appalled by the way her col-

***Mir*, of course, had already had women visitors, whose urine had to have been recycled along with that of the male cosmonauts. The instructor's explanation can only be interpreted as deliberate harassment.

league Norm Thagard seemed to accept their attitudes. Thagard, a pilot and a physician, had joined NASA two years ahead of her in 1978, in the first class that accepted women. He had been expected to treat women as partners, which he did—at least in Texas. Yet in Star City where Dunbar and Thagard had to spend at least eight hours a day together, either in a small classroom or mixing with cosmonauts, their relationship, never close enough to be deemed friendship, fell apart.

Dunbar could not laugh with the Russians. She flinched when a Russian put his arm around her, commenting that she "looked beautiful"; she would respond that her looks were not important, that what mattered was her Ph.D. When Thagard was there, as he often was, and these comments flew, she was hurt that he never came to her defense.

Eventually an escalating series of misunderstandings led to a total breakdown of communication between the astronauts. Dunbar and Thagard stopped talking while they were still training. Four years later Thagard explained his position to a visiting journalist: "It's not my job to defend her. You've got to understand. This is not the U.S. It's Russia. What am I supposed to do, take on the whole Russian culture?"

Dunbar did not get to fly as Thagard's backup, but she did spend five days docked to the Russian space station on the shuttle that retrieved him. She was still smarting from her Russian experience when she was interviewed later in Houston about her stint in Russia. She recalled: "I saw no women military officers at Star City. There were no female military lecturers." And the Russian women accepted, without comment, what she saw as subservient lives. Her resentment of Russian chauvinism strained her relations with her Russian hosts. But the responses of her male American colleagues to the Russians' behavior exposed fault lines in their attitudes toward women. In regard to Thagard's in particular, she said: "There are probably still some men at this Center [JSC] who believe it did not happen in Star City. That's unfortunate because all you have to do is ask even some of the cosmonauts, or ask the women in Russia, and they will confirm that there's still a real cultural challenge integrating women into their society equally. . . . Our challenge is to not change the way we do business for fear of offending the Russians. If we believe the way we're doing things is right—and it is—then we must stand up for it. There is no gender-related clause in the Constitution anymore. If we believe in that, we must stand up for it, which means the men in this Center need to stand up for the women if they're challenged by the Russians."

Thagard saw things differently. From his perspective, Dunbar was the ultimate ugly American trying to impose her American standards on another culture. "Bonnie just brought over to Russia all this feminist baggage that no one could understand."

She would not stop asking, why are there so few women cosmonauts?

Elena Kondakova responded independently and perhaps disingenuously, to the same question. "Because so few apply."

Dunbar did not relent: She felt that her determination proved that the Russians were educable, for she eventually won the respect of the cosmonauts. At what passed for a graduation ceremony, the head of training, Yuri Kargapolov, "got up . . . and he looked at me and gave me a toast. Now he had been the gentleman that had been hardest on me for the prior twelve months, and basically it was a toast for perseverance. General Dzhanibekov came up to me afterwards and said, 'You know, you American women are tough. We would like you to fly a long flight with us anytime.' So I felt good. I felt that regardless of anything else, I had done what I set out to do, which was to finish my job . . . and establish some relationships with the people there, and maybe they would think a little bit differently about women in their program."

Most probably she did gain a small victory. The Russians had learned to respect her. But what she did not understand was that they put American women, astronauts or anyone else, in a separate category. They still tended to ridicule their own ambitious women, and perpetuated myths about Valentina Tereshkova's space sickness in 1963 as emblematic of all Russian feminine behavior. They said they would welcome her, specifically, on a long flight, anytime. But Dunbar did not stay in Russia or fly with any Russians. She returned to Houston, leaving the classrooms in Star City to the next American team.

Meanwhile, in 1996, as Thagard and Dunbar trained in Star City, an infusion of dollars helped *Mir* grow. Soon Spectr, the new American-funded science module, linked up with *Mir.* The next year, Priroda, furnished with American laboratory instruments, linked up. Complete now with its solar panels extended like giant wings, *Mir* looked like a great dragonfly, stunning those who saw it against the blackness of space.

Inside, however, the aging space station, never luxurious to begin with, was filling up with an increasing accumulation of clutter. *Mir* was long past its life expectancy, and the resident cosmonauts had little time for scientific research for they spent most of their waking hours repairing the station's various support systems. Soon a series of quasi-catastrophic accidents would have them working overtime to keep the station habitable. For all that, *Mir* still served the Russians well, as a laboratory, observatory and a source of funds.

Thagard flew to *Mir* in a *Soyuz* in March 1995, the only astronaut to use the Russian "taxi," and four months later he was picked up by the *Atlantis,* with Dunbar in the crew. Eight months after his departure, the next American took his place. Shannon Lucid would be the first American woman to work on a space station and she would set a record for the number of days in orbit.

Lucid had had an even harder time in Moscow than Dunbar, but in different ways and for different reasons. She had not seemed perturbed, for instance, at a press conference before her flight when Yuri Glaskov, a top general in Star City, said that he was looking forward to Lucid's arrival on *Mir.* The place was a mess—everyone knew that—and he knew that "women love to clean."

Lucid probably didn't mind housework—in her own house. The oldest woman in the groundbreaking class of 1978, she is difficult to categorize. She had her career, but that didn't mean she rejected the traditional pleasures of a home. The only astronaut candidate to arrive in Houston with a husband who had left his job to be with her, she also brought along three children. She stood out physically because she was tall, and at five feet eleven inches she could fit into a man-sized space suit. Whether it was nature—she was born with an upbeat attitude—or nurture—she spent the first months of her life in a Japanese POW camp with missionary parents—Lucid seemed more grounded in family and faith than her classmates. Yet she was no less ambitious.

She loved to fly and loved exploration. Unlike her colleagues who had been pressed, cajoled and semi-bribed into going to Russia, Lucid had volunteered happily for the adventure. In 1996 she was fifty-three years old, her children were grown, and she was free to be away from home, home on Earth, for as long as four months—which is what she had agreed to. The last person in her astronaut class to fly, she leaped at the chance to go to *Mir:* "For a scientist who loves flying, what could be more exciting than working in a laboratory that hurtles around the Earth at 17,000 miles per hour?"

When Robert Hoot Gibson,[†] then head of the astronaut office, and also a member of the class of 1978, offered to send her to the language school in Monterey, California, to learn Russian, he had told her that after learning the basics she would be trained in Russia. She agreed to go even though he did not promise her an assignment. Sweetening the arrangement, however, he told her she would train with John Blaha, a friend and colleague with whom she had flown two missions; he would first serve as her backup and then follow her to *Mir.*

[†] Gibson is married to astronaut Rhea Seddon.

She arrived in Star City in January 1995 without much money, money she could have used to call home. NASA wasn't always generous with its astronauts, and there was no telephone in her rooms. Until her daughter set up an account for her on Compuserve and taught her how to use email—still relatively new in 1995—she had almost no contact with the family, friends and colleagues she had left behind in the United States. She was so out of the loop that she did not know what her husband meant when he emailed her in April "Your brother's OK." Her brother, a lawyer in Oklahoma City, had been known to drop by regularly at the Alfred P. Murrah Federal Building and she had been unaware of the terrorist attack that destroyed it and killed 168 people inside.

In addition to feeling isolated, Lucid suffered from her weakness in the language. The three months of intensive study at Monterey had not left her fluent, and she was expected to learn about *Mir* and the *Soyuz* from Russian-speaking instructors. Language was crucial here because she discovered that the training in Star City, unlike NASA, spent minimal training time with simulators or with the crew but taught everything from books and lectures. She had to learn to fly the *Soyuz* in case she needed to evacuate the station in an emergency. Yet when she tried to find out what else she could do later in helping with the station, she was told that her job was to be quiet and not interfere.

The other big problem was her isolation from NASA. During the first few months, she remembered, "There was no contact between NASA and us in Star City—nothing . . . I didn't even feel like I worked for NASA. I felt like I was working for myself. I never heard anything from the Phase One office," the people responsible for Shuttle–*Mir.* She and Blaha just kept studying and she assumed that NASA would tell her what experiments she would be performing in space. But that didn't happen. She had, it turned out, become the victim of a turf war between NASA contractors; no one would take responsibility for including her in the mission loop. Yet she pushed on. She worked hard and studied hard, and while frustrated with NASA's support network, she remained calm, most of the time.

While Dunbar had found Star City a bastion of "blatant male chauvinism," Lucid seemed either not to have noticed or not to have minded. She made friends despite her clumsy Russian and she took tourist trips to St. Petersburg and Moscow when her family visited. At this point, however, still in Russia and without help from NASA, she decided she would do her best, whatever the outcome. She "had had a good career, four shuttle flights, I was going to have a good time. That was it. I figured I would take care of myself."

But as her flight approached, Lucid grew tense. Eventually, at the end of January with the flight less than three months away, NASA sent Bill Gerstenmaier, the flight director, and Gaylen Johnson, her flight surgeon, to tell her in some detail about the experiments she would be expected to conduct. They were greeted by a woman who was totally uninformed about her mission, and they were amazed at her anger. Shannon Lucid, an astronaut with a reputation for almost saintly goodwill, "launched into an uncharacteristic tirade at the way NASA had mishandled her training." Then, with her fury released, she regained self-control. From that moment on she worked with this team, who remained in Russia coordinating her scientific activities with the people in the United States who had designed the experiments. They were available twenty-four hours a day, providing her with the support she had needed earlier.

A few weeks before launch, she realized that the Russians were nervous about sending another woman. They had sent four into space before her: Tereshkova, whom they considered a failure; Savitskaia, whom they had not liked very much; Sharman, who did not count because she was more of a tourist than a cosmonaut; and Kondakova, whose temperament some found trying.[‡] Suddenly they saw Lucid as an unknown quantity. They assaulted her with nitpicking questions, some surrounding her ability to use *Mir*'s toilet.

And then she was on her way. First she was off to Kazakhstan to watch her Russian crewmates leave on the *Soyuz,* and then almost immediately she headed back to Houston to train with the crew of the shuttle *Atlantis* STS-76, which would bring her to *Mir.*

There was not going to be enough time in Texas to get ready for months in orbit, so by email, Lucid asked her daughter to put together a small library she could bring with her. Her daughter, an English major, stocked it with eighteenth- and nineteenth-century classics. The library would prove a godsend and email would prove a lifeline throughout her mission, keeping her in touch with home and providing a way for her to keep a virtual diary. On March 22, 1996, the shuttle *Atlantis* lifted into the sky, to carry her to the rendezvous with *Mir.*

By the time she arrived at the space station, her two Russian crewmates had been there a month. Lucid recalls that "They acted very happy to see me.

[‡] Kondakova had become a star, a diva. She had her own celebrity as well as her status as Ryumin's wife. In contrast to her success there is the frustrated career of Nadezhda Kuzhelnaya, the only other female cosmonaut who has been awaiting her first flight since 2000. She was bumped from her crew assignment on the space station and then told that her arms were too short for the space suit and she would probably never fly. In Russia you are not officially a cosmonaut until you fly, and she may never earn her wings.

I believe they really were." She was equally pleased to see them—Yuri Onufriyenko, the commander, and Yuri Usachev, a civilian flight engineer, whom she took to referring to as the "two Yuris."

Onboard, Lucid took stock of her new home. *Mir* was filled with things that no one had been able to get rid of, including parts of old machines, tools for which there was no longer a use, and instructions for things that had been stowed away or sent back to Earth. It was filled with five more years of broken tools and equipment than Sharman had described in 1991. The base block, the center of the station, struck Lucid as about the size of a Winnebago—forty-three feet long and fourteen feet wide. It had two curtained sleeping nooks, a working compartment with computers, and a living compartment dominated by a red-tiled dining table. The Yuris offered her a private space in the Spectr module, where she stowed her English Bible (and found a Gideon Bible, in Russian, already there). She tucked her sleeping bag away and each night unrolled it, tethered it to a handrail, and floated in. As the Priroda module carrying the equipment for her scientific work had yet to arrive, Lucid decided to concentrate on human relationships. "Yuri and Yuri were absolutely fantastic to work with. We just lived every day as it came. We enjoyed working together and joking around together."

She had some time late in each day to reflect, and she thought about her childhood in the Texas panhandle. "I had spent a significant amount of time chasing wind-blown tumbleweeds across the prairie. Now I was in a vehicle that resembled a cosmic tumbleweed." She might have also reflected on the impact of capitalism on the Russian space program when both Yuris suited up for a spacewalk. They left the security of the space station with what turned out to be a giant Pepsi-Cola logo, which they then filmed against the cosmic background. (*Mir* was short of funds.)

Unlike the shuttle, *Mir* did not have constant contact with ground control. Instead, life on the station followed a ninety-minute cycle. When the station passed over Moscow, the crew had about ten minutes to catch up on news, directions and any help they might need. Lucid supplemented these rushed exchanges with email and occasional ham radio connections.

She adhered to a rigorous schedule, which she believed was the key to what was a very successful mission. She and the two Yuris ate all their meals together—including a late-night tea—and they exercised for forty-five minutes every day before lunch. She attached herself with a harness and bungee cords to one of the two treadmills and wore her new Penguin-3 "Muscle and Bone Loading Suit," made and manufactured in Russia. It was designed to lessen some of the debilitating effects of weightlessness, especially bone and muscle loss. The blue suit fit snugly, especially against her lower extremities.

Combined with regular, intense exercise, it may account for Lucid's fine muscle condition when she returned to Earth in September.

At least once a day she discussed her work with the team from NASA who had stayed in Star City to support her. She was expected to keep data on her own physical condition to go into NASA's data set. But this information has remained private, according to her wishes, seen by no one but her own physician. The data include the results of hours on a stationary bicycle and a treadmill. She hated the treadmill but kept at it daily nonetheless. Before zipping into her night sack she helped with the housekeeping, which was a constant struggle on *Mir,* and is by necessity obligatory in space, where crumbs can become missiles and the air fills with debris when least expected. Finally, before going to sleep, she read for about an hour, discovering that she liked Dickens, especially *David Copperfield.* She had brought a few books with her, mostly selections made by her daughter, and was frustrated to discover that she had only volume one of a science fiction series and had to wait weeks for volume two to arrive on the next supply ship.

Lucid's work life on *Mir* finally began when the Priroda module arrived with the scientific supplies. Once she had the experiments running, she coordinated her activities with the American and Canadian scientists who had proposed and designed them: She was responsible for carrying out whichever experiments had been selected by peer-review boards of biologists and physicists as well as chemists like herself. As soon as she could she set up racks of test tubes and performed the experiments she had practiced in Star City and Houston. She would return the data and samples to Earth on the space shuttle and they would be sent to the principal investigators of each experiment for analysis and publication. Lucid believes that her experience on *Mir* clearly shows the value of performing research on manned space stations. "During some of the experiments, I was able to observe subtle phenomena that a video or still camera would miss. Because I was familiar with the science in each experiment I could sometimes examine the results on the spot and modify the procedures as needed. . . . Only one of the twenty-eight experiments scheduled for my mission failed to yield results because of a breakdown in the equipment."

These experiments ranged from observing a controlled fire inside a box to monitoring the development of fertilized Japanese quail eggs (whose growth on *Mir* resulted in four times the abnormality rate of the control group of embryos on Earth). The wheat she had planted on her arrival in the Kristall greenhouse a few weeks earlier—which had delighted the crew as it began to grow—turned out, when examined some months later by experts at Utah State University, to have empty seed heads. None of these experiments were chemical, but all demanded experience with bench science, keen observation, scientific record-keeping and an educated, human eye that could note the unexpected.

Lucid never complained about having too little time to complete her work. On the contrary, she later cautioned those at NASA in charge of the International Space Station that it was imperative that scientists on long-duration missions have *enough* work to do and, equally important, that the work involve the astronaut's expertise. Too much free time, she warned, could weigh heavily on an astronaut. Her own experience was especially instructive because her departure from *Mir* was delayed six weeks. She had finished her assigned work before her original departure date and in the extra weeks that followed she had to fill the hours on her own.

By her account, besides reading, she spent a lot of time talking—in Russian—to the two Yuris. They probably remember her voice, she said. "We never spoke English. I mean, they didn't speak English. And I do like to talk, and I talk all the time, and I talked all the time when I was up there. They never said, 'Oh Shannon. Just be quiet. We can't stand to listen to your Russian anymore.'" When she didn't know the word in Russian, she simply added a Russian ending to an English word, which prompted one of the Yuris to comment: "You don't speak Russian. You have invented a new cosmic language."

She had heard about the disparaging remarks and chauvinist attitudes of Russian male cosmonauts, but she never heard or sensed anything from the Yuris—except once: She recalls she overheard them talking pejoratively about Russian women and she interrupted them, saying: "Now you guys, that's not the way you are. That's not how you've been treating me." And they explained that was because she was American. Because she came from such a different culture they likely did not expect her to behave in what they considered a feminine manner. She was an exotic creature, who they seemed to like, but she was not and would not be part of their lives on Earth.

When asked about her most cherished memory of the mission, Lucid said: "Well, I think when I look back, what I think most about is working with Yuri and Yuri and being with them. I think that makes the difference between a great flight or not. It's the people you work with. But that's true of everything in life. What you really remember are the interactions you have with people."

Except for the delayed arrival of the Priroda module, the mission went according to plan. It was also, unlike the Shuttle–*Mir* missions that followed, trouble-free. The glitch she lived through, the delay of the shuttle that brought her home, was an American glitch. The *Atlantis* had been scheduled to retrieve her in late July, but NASA engineers were studying burn patterns on the solid fuel boosters, which delayed its departure. When she was eventually retrieved, she had orbited for 188 days, shattering Kondakova's woman-in-space record.

When asked to comment, Kondakova congratulated Lucid, noting, "Life is easier onboard because you don't have to do laundry there, you don't have to cook there. So I think that, for a woman, being in space is kind of a vacation."

Lucid's own reaction to learning of the delay had nothing to do with escaping the laundry. She reports that she thought, "Oh no, another six weeks on the treadmill!" Then she focused on how she could make the most out of the extra time. One thing she did not give up was exercise, and when she finally returned she surprised the flight surgeons by walking, unaided, off *Atlantis.* Her physicians at NASA attributed her strong leg muscles to the dreaded daily treadmill workouts.

Lucid did not have any scientific duties to perform during the extra weeks and she must have had mixed feelings about the impending change of crew. According to the original schedule, she should have been gone when the Yuris left, but of course she wasn't. The Russians kept to their plan and sent the *Soyuz* with a fresh pair of cosmonauts to replace her friends. They also brought along a third party, another paying guest but one with more professional qualifications than Helen Sharman.

Claudie André-Deshays (now Claudie Haigneré), the fourth and final woman to visit *Mir,* left Baikonur on August 14, 1996, and arrived at the space station two days later. She had been sent by the French Space Agency, who called her a "spacionaut." France had its own space agency before the rest of Europe, and although it has become part of the European Space Agency, France continues to support its own research. France had enjoyed a close relationship with the Russians for some time and had sent French cosmonauts into space from Baikonur even as it was sending French astronauts into space on NASA's shuttles. The French paid Russia for using its facilities to train cosmonauts, starting with Jean-Loup Chretien, who flew as a cosmonaut in 1982. In exchange, the Russians provided office and some laboratory space for the French agency. In 1996, twelve years after Sally Ride became the first American woman in space, the French sent their first, and only, female "spacionaut" to train in Moscow and fly to *Mir.* Haigneré is still the only West European woman to have flown in space.

By any standard, Claudie Haigneré has remarkable credentials. A precocious student, she grew up in Le Creusot in central France, with an older sister, a younger brother and a thoughtful father who, as an engineer, traveled all over Europe and brought back souvenirs and stories, giving them a sense of wonder at the world. At fifteen she completed her "bac" (university entry examinations), then earned a medical degree and worked as a physician at Paris's venerable Cochin Hospital, continuing with a doctorate in neurology. She was also a gymnast, with a kind of Gallic charm that would win her an admiring public. Eventually she became a government minister, part of the French cabinet.

Haigneré had not planned to be an astronaut, but she cites the night in 1969 when, as she recalls, Neil Armstrong changed the world. She was twelve years old and from 9 P.M. through the early morning hours she watched the TV screen inside her home while outside, in the darkness, she heard celebrations. On the television screen she could see him—the first man on the Moon. And outside her window she saw the Moon as it had always been, only now she knew that it was not as it had always been because two men were walking across its surface. She was never the same.

A woman's place in France differs from that in either the United States or Russia. From the earliest years of the twentieth century, women have worked outside their homes with the government's blessing in the form of subsidized child care. Little girls in France learn about Marie Curie, and French women routinely practice medicine and become scientists. Claudie Haigneré believes that although the French system was not entirely egalitarian, the door to opportunity was open. "Maybe not fully open. But it was open."

When she saw a bulletin at the Cochin Hospital in Paris announcing a new selection of French astronauts who could be either pilots or medical doctors, she noted that the bulletin did not mention gender. So she applied. She thinks back to her decision and says: "I decided to try because, . . . maybe, because of the Moon."

She was one of a thousand applicants, a hundred of whom were women. She was the only woman selected. That was 1985, and she had already begun studying space-related medicine in anticipation of a new career. Seven years later she went to train in Star City, where she became the backup to another French spacionaut, Jean-Pierre Haigneré, whom she later married.

She liked living in Russia, which she did not find as inconvenient as NASA's astronauts did. She was nervous at first, and "thought at the beginning it would be difficult to cope with Russian men." Fearing, perhaps, they would be disdainful, but she encountered no problem there. She was training as a commander of the science system, a high rank which the Russians respected. "There is no [other] woman in the world who has that qualification. I was not a man or a woman. I was a working cosmonaut and I had to be prepared to perform the job."

It was teamwork, but also a kind of chivalry. She was a cosmonaut, of course, but she was also French, a culture Russians had romanticized during their Czarist past, and admire still in the capitalist present. She was also a guest whose country was paying more than 80 million francs ($16 million dollars) for her to spend two weeks doing research on *Mir.*

In sending "spacionauts" into orbit, France was declaring that it was a player in the adventure of space travel and research. In sending Haigneré, it was proclaiming its pride in French women. Russia, by its partnership with France, was also making a statement about its commonality with France.

Haigneré left Kazakhstan in a *Soyuz* along with two replacement cosmonauts for *Mir* on August 14, 1996. Two days later she boarded the space station and was greeted by Shannon Lucid.

Her first impression after leaving the cramped *Soyuz* was that *Mir* seemed a vast labyrinth. Her hosts led her directly to base block for the traditional welcoming ritual of bread and salt, and then they feasted on the fresh cucumbers, tomatoes, lemons, potatoes and cheese she had brought with her. Haigneré recalls how good the fresh food smelled in the station's aseptic atmosphere. She realized immediately that scents and flavors are often weaker or stronger but not predictable in space, so she did not use the perfume she had brought with her.

She knew that an American astronaut would be working there, but was surprised by the warm reception. "Shannon had prepared for my arrival. She provided my bed, a place for my books." She also showed her how to use the dry shampoo and she lent her a recorder to listen to music. After four months, this was Lucid's home and she behaved like a hostess. According to the Frenchwoman, "Shannon was a mother to us all."

Haigneré recalls how difficult it was orienting herself the first day, how finding things was hard because *Mir* was so big. There was one module she had to enter on her stomach, and another on her back, and yet a third she had to enter on her side, to fit in where it seemed she wanted to be. While she learned all this, she seemed to get lost all the time.

Haigneré credits Lucid for easing her way. "It's this kind of attention that was so nice. I think it's mainly a feminine attention." Rethinking her praise she qualifies her comment. "It was mainly Shannon. I'm not sure that all women are that way. I remember too, the day before landing. There were a lot of things to do, to pack. Shannon saw that we were very busy and we had no time to think about little details. 'No problem, do what you have to do. I will do it for you.' It was hard to leave Shannon."

Haigneré had been responsible for the space medicine and physiology programs at CNES (Centre National d'Études Spatiales—the French Space Agency) and on *Mir* carried out a series of experiments. She had brought along a portable laboratory called "Physiolab" that she used to track the astronaut's blood pressure and heartbeat as well as the imbalance that some astronauts experience when they turn their heads. In another experiment called "Cognilab," she measured how people orient themselves and recognize each other in the disorienting conditions of zero gravity. A different set of experiments involved injecting hormones into female salamanders (an animal, the French press reminded its readers, that was the emblem of Francis the First). The species Haigneré used mate early and the females keep the sperm in a special sac and so are able, when triggered by hormones, to produce fertilized eggs. On board *Mir*, according to a colleague at the French space center in Toulouse, there were "enough eggs to almost make caviar." The experiment focused on the developing embryos as well as the impact of microgravity on the question of fertility. The embryos did develop normally, proving that amphibians can reproduce in zero gravity. Whether they will want to has yet to be determined.

Acting in much the same way as Lucid, that is, as a scientific generalist, Haigneré functioned on *Mir* as an entire scientific faculty—biologist, physician and physicist. For example, in addition to experiments on herself, for physiological data, and with the salamander, she performed a physics experiment called "Alice" to observe the actions of fluids at the critical point when they become gas.

She had nonscientific responsibilities as well, including hosting a French dinner prepared by students at a hotel school in Souillac. The French wanted to remind the world of their "culinary tradition," and tried to modify their recipes "to the exacting specifications required to feed astronauts in microgravity." The menu included daube de boeuf, confit de canard with capers, pigeons in wine and tomato confit, all washed down with wine from Alsace. The French newspapers reported every bite.

When *Mir* passed over France, Haigneré chatted with children and adults about life in orbit, answering questions about her time in space, especially about the mix of French, Russian and American cuisine. It was a high point for the French mission. Lucid never mentioned the meal.

While dining in orbit, Haigneré discovered that the crew had a tacit understanding about space etiquette. For example, "When we wished to leave the table, we moved around the table and we never passed on top of the table. It would not have been polite."

Haigneré herself is very polite and even diplomatic in her effort to please all her benefactors. Yes, she says, all-male crews on space stations have been successful in their missions, but the success of the mission is not the only thing that matters. She feels that Europe ought to have sent at least one woman into orbit "because we are a mixed society where there are men and women." She asks quite logically, "So why are there no women in space?" She speaks of the importance to womankind, a way for the women of Europe, and perhaps the rest of the world, to feel represented. She also uses the term "complementarity," by which she means that women and men have qualities that enhance each other. And thinking practically about a possible trip to Mars, or a settlement on the Moon, she says: "I'm sure that there will be babies on the Moon. There will be life, as on Earth. And that, I suppose, means that there will have to be women and men in the spacecraft on those expeditions."

In a blink, Haigneré's trip was over and she returned to Earth with the two Yuris. In the next weeks she enjoyed a hero's welcome in France while Shannon Lucid continued orbiting on *Mir.* Finally, on September 22, the shuttle *Atlantis* arrived with John Blaha, the astronaut with whom she had trained in Russia. She explained to him what she had been doing during the extended changeover period, and they switched places. Shuttle–*Mir* would continue for five more crew exchanges.

There are critics who believe that NASA wasted valuable resources, as well as time, collaborating with Russia. It is possible that the International Space Station could have gone up sooner had NASA not spent time and resources sending astronauts to *Mir* to learn about living on a space station. But by going to *Mir* Americans learned how a space station works. They also learned, especially from Shannon Lucid's lengthy stint, the importance of paying attention to the psychological and cultural composition of the crew.

Meanwhile, in September 1996, a third American woman astronaut arrived at Star City for training. Wendy Lawrence, an Annapolis graduate, helicopter pilot and physics instructor at the Naval Academy, had joined the astronaut corps in 1992, completed training in September 1993, and by December, in record time, had been assigned to a shuttle flight—STS-67, a science mission that in March 1995 had set a duration record of sixteen and a half days. She was eager for another flight, and to raise her chances she volunteered to go to *Mir* and was accepted as John Blaha's backup and eventual replacement. Before she left for Russia, NASA had understood that Lawrence just made the Russian minimum height requirement of 160 centimeters. But four days before she left Lawrence learned of a memo saying that the minimum height was 164 centimeters. What to do? Since she already had a ticket, NASA sent her anyway.

She worked in NASA's office there and returned to Houston to support Lucid. In March, Robert Cabana, the new head of the astronaut office, asked her to return to Russia as the director of operations. And while she was there, they arranged for her to get measured again, to see if she could fly in the *Soyuz.* She fit into the spacecraft, but her arms were too short to fit into a Russian space suit, the one that they use for spacewalks. They tried their best, first cinching it up as much as they could, then pressurizing it, but she couldn't get her hands into the gloves. There was, in fact, so much room in the suit that she could take her entire arm out of the glove, and out of the arm of the suit, to scratch her legs. It wasn't going to work. She told the Russians that it wasn't safe because she could put both hands inside the gloves. And they wrote a letter saying she could start training if she agreed never to be considered for a spacewalk. She could not agree to that.

The American press expected her to be angry, or at least unhappy, about being replaced by a taller, male astronaut. She wasn't, however, because what the press did not know was that she was rewarded for her decision with an assignment to fly on STS-86, the flight that was supposed to have taken her up to *Mir* and left her there. While on STS-86, Lawrence visited *Mir* as they transferred equipment and another astronaut, Dave Wolf, to the space station. Two months later, she learned that NASA needed a Russian-speaking flight engineer for STS-91, the last flight to *Mir*. She got to *Mir* twice, and was followed by three other female American astronauts who, like Lawrence, helped supply the space station although they were never residents.

All of the women *Mir* hosted enjoyed the experience, but it was especially exhilarating for Shannon Lucid. Her mission bore every earmark of success. Her Russian colleagues sang her praises and her name still elicits smiles in Star City. She won praise in Houston for her adherence to the exercise program and her scientific duties. After her return, having set a record for female endurance in space, she celebrated with President Bill Clinton at the White House.

But there were problems. When she turned her assignment on *Mir* over to John Blaha, her backup, her colleague, neighbor and one-time close friend in September 1996, things went badly. Blaha felt that Lucid didn't communicate clearly during the exchange period. As a result, he failed to understand the work she had accomplished and was generally unprepared for life with the Russians on the space station. Where Lucid described the time she had to relax and read, Blaha spoke of working without stop, of never really feeling on top of things, although his workday included the same down hour and reading time as hers. He kept notes on everything to help future astronauts. He saw the glass half empty whereas she saw it half full. Their reaction to their missions could not have been more different.

Blaha seethed as time passed and he grew increasingly angry at Lucid. "I will never communicate with her again," he said. In his eyes, she had been his "best friend," and she let him down. He retired from the astronaut corps while Lucid remained, and in 2002, she moved to headquarters in Washington as head of NASA's Science Office. In the nineties, on *Mir* there was no gender discrimination between Russian men and foreign women as there had been among Russians. There had been friction a decade earlier between Svetlana Savitskaia and the men in her crew, and there was some discord between Elena Kondakova and her crew. They treated their own women one way, and foreign women, whose countries had paid for the trip, as anyone would treat a valued client.

At home in Star City, American men and women noticed a cultural divide between Russian men and Russian women that reminded some of them of the way Americans had behaved a generation earlier. When NASA first accepted women into the astronaut corps in 1978, the agency went to great lengths to eliminate gender bias. Many astronauts, women and men, believed that it had worked perfectly. Norm Thagard seemed to sympathize with Russian men about the proper, subservient role of women—or at best, he seemed to be reluctant to criticize their attitudes. His resentment festered, exploding when he felt slighted during his post-Mir visit to the White House where he did not get the major reception accorded to Lucid. "If Clinton knew who I was when I visited the White House, he gave no indication. But he was at the landing for Shannon. He just didn't have much use for a white Anglo-Saxon male. But for Shannon? A woman? It was night and day."

Dunbar accepted a certain amount of raw chauvinism from the Russians, but she was distressed when she found it in an American colleague. The ugly voice of bigotry from Thagard, a former Houston buddy, revived old memories of discrimination, of having been denied a recommendation to engineering school after scoring higher on her SATs than her boyfriend who received an application form, because there was no place on the form for girls who wanted to become engineers.

Wendy Lawrence saw things differently. When she was director of operations in Star City she worked in an office with three Russian women whom she calls "our secretaries/troubadours." She had a different impression of gender relations from them. "I got to listen to them talk about Russian men and, in reality, Russian women really do run the country. They just let the men continue to think that they do. . . . They were all really spunky women, very capable women, [who] very much held their own and wouldn't let any of the men boss them around. . . . The real huge difference for me is I had already flown. The Russians have an incredible amount of respect for cosmonauts, for anyone who has flown." And that made her's a special case. She was treated as a hero from the moment she arrived in Star City because she was a flown astronaut.

Perhaps that explains her impressions of Russian life, especially the lives of Russian women. The idea that women *really* run everything is popular in countries where women have little real power. The "power behind the throne" is a theme as old as civilization, and in some instances may be true. But this illusion of power does not get Russian women into space, or into high political office, and has seldom helped divorcing women win child support in the event that the throne has chosen a different, perhaps younger, woman to wield power.

During the last third of the twentieth century, women in Western Europe and the United States steadily gained access to educational opportunities that had been closed to them, and to jobs outside their homes that had previously been off-limits. Russia faced a different set of struggles, and gender equality was not high on the agenda, in part because women already studied mathematics, science and engineering. Bonnie Dunbar chafed at what she took as insults because they suggested that the changes she had lived through in American society, changes that had allowed her to become an astronaut, were not universal. What a man like Norm Thagard accepted cheerfully as adjusting to a different culture could be interpreted as indulging a certain nostalgia for a time when men like himself—and only men—could flaunt their medals and luxuriate in having "the right stuff." He was not the only American man whose enthusiastic acceptance of female colleagues masked a longing for the old days. John Blaha's rupture with Shannon Lucid revealed that another man in the astronaut corps resented sharing space with a woman.

On the bright side, the non-Russian women who orbited on *Mir* succeeded with most of their scientific tasks and, as important, were able to communicate professionally and personally with Russian colleagues. They were able to live comfortably with men from a different culture who spoke a different language. The women were all civilians, the Russians were all military men, and that difference in outlook did not interfere with the operation of the station. But the civilian–military divide had a different cast in the American astronaut corps. As more military women joined NASA, new leadership roles would reveal different ways of exercising authority.

10

OFFICERS AND GENTLEWOMEN

EILEEN COLLINS BROKE THE FINAL BARRIER in the astronaut corps when she was accepted as a pilot astronaut in 1990. That class marked a turning point. Collins was the first female pilot astronaut, but two of the five other women in her class of twenty-three were also from the military. NASA's pool of female recruits had come a long way since 1977. Back then it had advertised its new gender-neutral policy in universities, medical schools and women's scientific organizations. In 1990, NASA was finding many female candidates among its own employees at NASA centers in Cape Canaveral, Houston, Huntsville, Mountain View and Pasadena. It was also attracting women from the Navy and Air Force.

NASA, from its inception a civilian agency, had always been conflicted about its military roots. In the beginning, of course, almost all the astronauts were military test pilots, as was most of the corps right up through Apollo. But that was only half of it. Many of NASA's administrators and engineers also had military backgrounds, although some of them became civilians when they decided to remain at NASA. They brought with them an unmistakably military language—astronauts go on *missions*—and a chain of command style of managment that extends to the conduct of science in space.

With the integration of an increasingly large number of women into the armed services, the astronaut selection committee had a chance to choose military women, and it did. After *Challenger* some of the old hands wondered if the civilian mission specialists had ever, in their guts, understood the uncertainties of space travel. When the military asked that more of its own be accepted into the astronaut corps, NASA obliged. The agency was comfortable with women who had a soldier's understanding of danger and were used to both giving and taking orders. When NASA began accepting more military women in 1990 it is likely that they expected these women to be just like the guys. They were in for a surprise.

NASA never changed its original requirement that they preferred that their pilot astronauts had military test pilot training. What changed was the military. Recall that at the beginning of the human space program, women were not admitted into any military flight training programs. Finally, in 1974, the Navy accepted women for pilot training, and the Air Force followed two years later. But it was only in 1988 that the Air Force allowed women into their test flight training program. At last women like Collins had a chance. At test flight training she was one of only four women at Vance Air Force Base in Enid, Oklahoma, among 800 male pilots. It was hard, but she had long before decided that she would be a pilot astronaut, and went with the flow.

To the public, pilots were still the *real* astronauts, the guys at the controls who had been described as skillful daredevils with the "right stuff." By the time Collins applied, the "right stuff," a phrase coined by the writer Tom Wolfe in a 1983 series in Rolling Stone and later as the title of his 1979 account of the Mercury Seven, had become something of a joke in Houston. Although Collins was both brave and skilled, she was neither foolhardy nor immature. She was a lieutenant in the Air Force—self-confident, sensitive, compassionate, and a force to be reckoned with.

Collins joined NASA in 1990. By the time, five years later, when she became the first woman to pilot a shuttle, NASA had accepted two more women pilots, as well as three women officers with advanced degrees in science and engineering as mission specialists. These military women, both pilots and mission specialists, had, in a sense, proved their strength and brought an unembarrassed sense of themselves as women to the astronaut corps.

The women altered the tone of the astronaut office, the engine that runs the human space program. They arrived with a special kind of experience. Whether Air Force or Navy, they had been initiated into what had been a "brotherhood." Although the Air Force culture differs from the Navy's, both institutions oblige its women to adjust to what the women call "a man's world." That process can leave a searing impression.

In Houston, Collins learned right away that she and her male counterparts would have a separate career trajectory from the mission specialists, a fact that underscores an inherent inequality in the system. When the second and third women pilots arrived in the class of 1995, they noted this too. Pamela Melroy, Air Force, and Susan Kilrain, Navy, knew that as pilots they were in line to become commanders, which meant that they would be in a position to help select their own crew. And in orbit, where there is a strict order of command, they would be on top.

Melroy explained: "We women pilots are in a little bit of a unique situation. There is a difference between pilots and mission specialists in the respect that a pilot is a future commander. What that means is that you're potentially going to be in charge. Maybe even be a potential chief of the astronaut office. You have a lot of impact on whom you pick for your flight . . . the mission commander will sit down with the chief of the office. The chief of the office will say, 'Here are the people I think need to be in your flight.' But if the mission commander says, 'I'm not taking that one,' they won't. Inevitably pilots have a different relationship with mission specialists than the mission specialists have with each other. There will come a time when they will be in a position to affect the careers of other astronauts."

These women knew that they were potential heads of the astronaut office, and whether they were Air Force, like Collins and Melroy, or Navy, like Kilrain, they suspected that as leaders they would be different from male leaders. Certainly the Air Force and Navy had treated them differently from the men, and seldom gently.

Whatever the root cause, most of the military women in the astronaut corps, pilots and military mission specialists alike, share attitudes that distinguish them from their male colleagues and often from the civilian female mission specialists as well. The number of military women in the astronaut corps grew regularly after 1992, with about two military women entering with every new class. That meant about half of the newer women, many of whom have never flown, had military training.

The first, Collins, piloted her first shuttle flight in 1995. Then Susan Still (soon to marry and become Susan Kilrain) and Pamela Ann Melroy joined the corps. In one fell swoop there were three female pilots. NASA had followed the same tactic as it had in 1978 when it took six female mission specialists. Once again it was clear that Eileen Collins could not be a token. All three pilots brought a strong commitment to the military traditions of expertise and discipline that had dominated the civilian space program from the start. Yet their image of military culture was a world away from the wild get-drunk-as-a-dog jock who took risks above and beyond necessity, a kind of machismo that was already fading in the astronaut office, but is still endemic throughout the military.

The entry of women pilots would mean more to the astronaut office than simply extending the applicant pool. Everyone knew Collins was an excellent aviator. What no one could predict was her leadership style. Would she lead differently from her male colleagues when she took the helm of the *Columbia*? And if she did, would that mean that all women pilots would lead differently from men? These are old questions and there have been many studies, including some done by the Air Force, contrasting leadership behavior among women and men. They have concluded that women do tend to lead differently, but just as effectively as men. The differences may be a result of socialization, personal experience or even some essential biological difference that we do not understand. They may also be a response to a changing world.

The successes women have had in business, politics and the military during the last forty years have changed how women function as part of the working world. The way women lead is probably different in the first decade of the twenty-first century from what it was in the 1950s.

In a study of women and the use of military force, the authors delineate two ways women in powerful positions operate–through cooperation and segregation. The first "tend to protect themselves by adapting the attitudes of their male colleagues. They 'go native in order to survive.'" The second "cultivate female friendship, support, and cooperation in order to cope with low status." Both of these approaches limit women "from having an independent effect on the structure and culture of mainstream organizations."

But this is a blanket description. Breaking the issue down into generations, Judy Rosener at the University of California at Irvine sees the older women conforming to male standards while younger women draw "on the skills and attitudes they developed from their shared experiences as women." Overall, her study suggests, military women usually adapt rather than segregate because there is so much misogyny, as exemplified in episodes of unpunished violence toward high-ranking or aspiring female officers in the Navy and at the Air Force Academy, where adaptation, suffering indignities and even outrageous insults "like a man," seems safer.

That said, NASA's female pilots are extremely sensitive to their role, their choices, and find the astronaut corps deliberately and self-consciously accepting of their presence. The astronaut corps is not the Navy or the Air Force. It is a unique organization that has been making an effort to integrate women for twenty years and, on the whole, with reasonable success.

This is reflected in studies that show when given a chance, women lead by consensus. Successful women seem to have an abundance of empathy. But whether this is merely a cultural phenomenon that will change as more women win more positions of leadership is impossible to predict. It is also true that many astronauts find these same qualities—empathy and leadership through consensus—in the male pilots they admire most.

Pamela Melroy has ruminated over the leadership styles of the other two women pilots and placed the three of them on a curve, identifying herself as "an intermediate woman."

To one side she placed Collins, who she considers a natural nurturer (Collins now has two children) who approaches people enthusiastically and with an apparent lack of guile. Cady (Catherine) Coleman, an Air Force mission specialist who was in Collins' *Columbia* crew, said that "the biggest difference [between Collins and the male commanders] was that Eileen had a different leadership style, a style that might be thought of as typically female—that is, more consensus-based as opposed to command-based."

Yet her gentle manner did not prevent her from mastering the operation of the spacecraft. Coleman continued: "How she handles things in maybe a more consensus-based way isn't connected to following the procedures to get yourself out of trouble when things are hot and heavy. It has to do with, 'Does everybody want to eat one meal a day together? Should we make sure that it's on the schedule?' Instead of, 'We're eating one meal a day together.'"

When it came to mastery, Collins was as macho as any guy. She did not flinch when a few minutes into her flight the computer indicated two separate problems—a hydrogen fuel leak and a loose wire causing intermittent shorts. When people asked if she was nervous or afraid, she replied, "Absolutely not." She had been through it in the simulators many times and she knew exactly what to do. "Do you know what makes you afraid?" she asked rhetorically shortly after her second flight. "It's not understanding and not knowing. When you know the system and you know the people that work on it, and you know how it works, you have control." That is the way the first female shuttle commander operated.* Collins established one model of feminine leadership.

Melroy sees the women as different in style, but not method, from their male counterparts. At the other extreme from Collins she placed Susan Kilrain, a lieutenant commander in the Navy. Kilrain is forthright, a no-nonsense woman who, as the only sister among brothers, learned as a child to stand up for herself. Kilrain did not think about becoming an astronaut until she had finished college because for most of her life, she just wanted to fly. She earned a pilot's license and instrument rating when she was studying

---

* In interviews following the *Columbia* tragedy in February 2003, Collins, the scheduled commander of the next shuttle mission, said that she was ready to go, pending a completed investigation into the cause of the tragedy.

engineering at Georgia Tech near her home in Atlanta. After college she got a master's degree in engineering while she worked for Lockheed, but she didn't really like it. When her boss, a good friend of *Challenger* commander Dick Scobee, put her in touch with him, Scobee suggested she join the Navy to become a test pilot. She did, and entered the Naval Test Pilot School in Patuxent, Maryland, where she flew F-14s, taking off and landing from aircraft carriers.

She was not naive and knew that the Navy was having a hard time adjusting to women almost everywhere, including its continued refusal to include women in its submarine crews. The Navy did not accept women as flight officers until 1979, although they had trained them as pilots beginning in 1974. When Kilrain enlisted as a Navy pilot, there weren't many other women and the only camaraderie she knew was among the men in her class. Despite the overall cool reception from her fellow pilots—which she had expected—Kilrain qualified as a naval aviator in 1987.

The Navy was not a very female-friendly organization then, but to get through Kilrain simply tuned out what she didn't want to hear. "I realized that I was a woman in a man's world, so I was going to be an outsider. My whole philosophy was not to make waves. My goal was to be an astronaut. I wanted to fit in without accepting unacceptable behavior. I felt like I did a pretty good job. At times there were people [in the Navy] that did not think women should be doing the job and [there were] men that were very outspoken against women pilots." She just ignored them and kept her eye on her goal.

She was already flying fighter jets when she applied to the astronaut corps in 1992. A year earlier, an infamous incident of sexual harassment occurred at the annual meeting of the Tailhook Association, a Navy officer's club, at a Las Vegas hotel. Kilrain, a member of Tailhook by virtue of having landed on an aircraft carrier, was there. During the days she attended sessions and met with admirals. During the evenings she stayed in her room. On one of those evenings another female officer, Lieutenant Paula Coughlin, had gotten off the elevator at the wrong floor, where she was trapped and manhandled by a mob of drunken officers. After the melee Coughlin had trouble identifying the assailants well enough to justify a naval hearing, and a year later, frustrated with the Navy's inaction, she told television commentators and the *Washington Post* that the Navy had failed to fully investigate the assault on a female officer. The Navy had no choice then but to pursue her charges, as the whole world heard what had happened. Within the Navy, Kilrain said, male–female relations suffered for over a year and all promotions, including the women's, were temporarily held up. In short, she entered the Navy at a time of turmoil. But in the wake of Tailhook, Kilrain explained, the Navy really tried to restore morale and improve gender relations.

She witnessed sexist incidents, but she never had any personal problems. She explains what happened in terms of the Navy's historical role, of which she is very proud. As the oldest service the Navy is heavy with tradition, including a reluctance to accept women. She knew that not everyone welcomed her into what had only recently been a male bastion, but she also noticed that the most negative responses to her presence came from poor aviators, men who felt threatened by her skill. The good aviators, she said, didn't care.

She describes herself as a natural flier, and notes that she "usually finished first in my class every step of the way." But that would not have been enough, she acknowledges, without "a lot of very brave women that have done a lot of fighting to open doors for me. And in that respect I feel very fortunate." Like some of the women who preceded her in the astronaut corps, she concedes that she was "never one to be out there willing to risk myself and my career. I was fortunate that other women did it for me. I wouldn't have been able to be a Navy pilot or an astronaut had these women not come before me. Somebody was willing to force the issue and break down the barriers. Definitely. I recognize it. I hope that in my own small way, just by doing it and being successful at what I've done, it helps future generations."

Kilrain and Melroy are the same age and both arrived in Houston on the brink of marriage to fellow servicemen and, not surprisingly, they became close friends. As it happened, Kilrain became the first pilot in their class to get an assignment, a mission that split in two parts.[†] That mission began as STS-83, the shuttle *Columbia,* and left Cape Canaveral on April 4, 1997, carrying the Microgravity Science Laboratory, a self-contained cylindrical module built by the European Spacelab Agency. But after three days in orbit it had to return because of a faulty fuel cell. Just three months later Kilrain got to relive the excitement and tension of a launch and reentry when the same mission, now renumbered STS-94, returned to orbit with the same crew working in the same laboratory module. The second time they orbited for sixteen days while mission specialists studied how different materials responded in zero gravity. The laboratory onboard carried three kinds of experiments: combustion science, material science and protein crystal growth. Kilrain discovered that on a Spacelab science mission the pilot is often under-tasked, which in her case meant that in addition to carrying out

---

[†] The split mission, unique in shuttle history, earned her only credit for one mission in terms of the two missions a pilot needs to qualify as a commander.

her piloting responsibilities, she got assigned to be the primary mission photographer for Earth observations.[‡] So while the scientist specialists had every moment scheduled, she was obliged to look out the window and "enjoy the awe of it all."

Pam Melroy is less accepting of the way things are than her friend Susan Kilrain. She is "so frustrated that feminism has turned into a dirty word. It certainly is hard to sit and listen. . . . Once you get acceptance you're one of the crowd and it's not a problem. . . . [But] I know how those civilians feel, how strange this is. . . . As a woman I can still straddle the cultures, because I'm enough of an outsider in the military that I can still look at it that way." Yet she grew up in a military family and is emotionally committed to the Air Force, an institution she obviously respects.

But she knows its quirks. When she was a child her Air Force father was caught in a version of Catch-22. After earning a master's degree in computer science in the Air Force in 1961, he became a computer expert when computers were still new, and "got exposed to a lot of top secret stuff." Seven years later, that experience caught up with him. The Air Force passed over him for a promotion because he had not gone to Vietnam. He had not gone because he *could not* go. He was forbidden to leave the continental United States because of the top secret stuff he knew. But, Melroy explains, he wasn't angry. As soon as he left the Air Force he had fifty job offers and eventually went to work for Xerox, in Rochester, New York. And that is where she grew up, sandwiched between two brothers.

Her mother, who did not have a career, nonetheless served as a model because she was "a born general." She was also "a good Catholic girl and military wife," who in those days did not work. She did have a business degree, so as soon as Melroy's youngest brother finished high school, her mother returned to work as a temp and before long, her proud daughter says, she had cornered the market for vice-presidential and presidential support office temps in Rochester.

On her own, with no special prodding but inspired by the Apollo astronauts, Melroy decided when she was eleven years old that she would be an astronaut, too. Like most adolescents, she had grappled with questions about the purpose of life and had decided: "Whatever I do, it needs to be something that I really believe in." She considers space exploration a kind of service industry. She realizes that she could have focused all that energy on

---

[‡] All astronauts are taught to handle a variety of photographic equipment in order to record the expected, and the unexpected, during their missions.

becoming a doctor, but she liked astronomy, not anatomy, and the idea of discovering something new. She reflects that the women's movement and some of the inequities she experienced were "just part of the world I grew up in. I had a little problem in high school with science in particular, they were just dumbing it down [for girls]. There were other things. People laughed at me my whole life when I said I wanted to be a pilot and an astronaut. . . . If anybody laughed at me about the way I talked, the way I looked, whatever, I would have fallen apart just like any other high-schooler would have. But for some reason I was shot out of the cannon and there was no doubt and there was no hesitation and I was very confident and very strong."

Her father encouraged her unstintingly. "In fact," says Melroy, "when I turned eighteen, I went to Wellesley [College and] he took out a life insurance policy for me." He told her that when she became an astronaut, she might have trouble getting life insurance because of the hazards of the job. She was touched. "You can't get much more supportive than that." At Wellesley, near Boston, she joined the Air Force ROTC that met at MIT twice a week and became the executive commander of her detachment. That gave her ten hours of piloting time, just enough to let her fly solo in her senior year. After that she trained on jets and spent a year at MIT getting a master's degree in astronomy with a theoretical thesis on the atmosphere of Neptune. (When Neptune's atmosphere was actually analyzed a few years later, her theoretical numbers matched the actual data.)

Up to this point, life had been kind to Melroy. Then she entered the Air Force. As a cadet commander in ROTC she had been in a unit that was about 40 percent female. Suddenly, in pilot training she was the only woman in the class. She quickly discovered that inordinate teasing was part of the military culture. The more they liked you, the harder they tried to be tough on you. In short, she was miserable. At home her brothers had never treated her that way. No one had ever insulted her dignity. It took her two years to get used to it, but she learned what it took to survive in that military culture.

She was only the third woman test pilot in the Air Force, and the second at Edwards Air Force Base in California, the legendary landing strip to which Chuck Yeager returned from his record making flight when he broke the sound barrier in 1947. There was no place for her to hide: "If there was a chick on the radio, it was me, and everybody knew who it was. If I made a stupid radio call, it's like 2,000 people on that base knew exactly who said it. There is a difference [in the way women were treated]. At Edwards, . . . my appearance worked against me." She mulled it over, considered trying to look less feminine, then rejected the idea.

She figured that the first woman to do anything plays a crucial role, and she was determined to do it well. If she tried to conform to something other than who she was, she would be "doing other women a disservice and doing it to myself." As the third woman test pilot, she had benefited from the two who had preceded her because she knew that all it took was for that first woman to be under par, or be disliked, and that would be it. The tone would be set. When you are one of the first, she explained, "people are going to have a mind-set about the other women who come after you. They're going to expect them to be like you. They're going to treat them instinctively the way they treated you. In that sense, yes, it is really important who is first and who is second and who is third. It makes a real difference."

She knows she is more like Amelia Earhart than one of the FLATS. Earhart was allowed to fly, only she was not accepted in all the races and she had to open a few doors. The FLATS wanted to be astronauts, but there was simply no way that a female aviator could get military test pilot training in those years. Melroy did not have to surmount an impossible barrier, but becoming one of the first pilot astronauts was a challenge and she, like Kilrain, hopes that they are making it easier for the women who follow them.

Of the three, only Melroy, a lieutenant colonel in the Air Force, has flown in battle. The first time was in 1990 during Operation Just Cause, when she flew without charts in bad weather with an inexperienced crew at a low altitude through a mountainous area in Panama—and barely missed slamming into a mountain. The next was Operation Desert Storm in Iraq in 1991. After these trials, she was looking forward to her first spaceflight. It would be the last for a member of the class of 1995, and the most complicated mission to date. STS-92 would carry and install a score of crucial parts to the then incomplete International Space Station. When the mission ended, the Station would be ready for its first crew.

Melroy describes how she rehearsed the mission endlessly in the shuttle simulators. She explains how no pilot can ever know in advance exactly what it is like to maneuver the shuttle in space. The simulator, an exact replica of the flight deck with all the controls, is the nearest thing. It provides the pilot with a variety of scenarios of where things can go wrong, which the pilot learns to fix. To get a more tactile sense of what it is like to work in space, she practiced moving and carrying things with the rest of her crew in the enormous Sonny Carter Neutral Buoyancy Laboratory. This pool is 202 feet long, 102 feet wide, 40 feet deep and holds 6.2 million gallons of water. Astronauts spend as long as eight hours at a stretch submerged in their space suits practicing EVAs—space walks—in the only environment on Earth that is anything like microgravity.

In May 1999 Melroy had debriefed Kilrain and listened to the seasoned commanders describe their responsibilities and impressions of spaceflight. But she still had not flown. As she thought about how and why she became an astronaut, she wondered if being in orbit would change her view of space.

Melroy never doubted that women differ from men. She knows that women can make it today in "a man's world," but she believes they make it their own way. She describes herself as the third *woman* test pilot in the Air Force. Unlike Judy Resnik, she never says that she is just an astronaut, not a woman astronaut. She is the third woman test pilot in the Air Force and the third woman to pilot a shuttle.

The differences between men and women pilots, she explained, are not in piloting skills but in personal concerns. For example, her friend Susan Kilrain had an unfortunate haircut—her bangs were trimmed really badly. "We have the same hairdresser and she called me up in tears and said in quarantine, 'Can you see if you can get our hairdresser medical approval.' When you're in quarantine nobody can see you without a physical. So I spent most of the day running it up the chain and defending that position to the senior guy over in headquarters saying, 'Look, she's going to have this picture taken of herself so she needs this.' They said all right. So I went and collected him, he canceled all his appointments for the day. He thought it was great. All you have to come back to are the pictures and that tape. That's all you have to remember this wonderful experience. Every time you look at the picture of yourself you think, 'I look like shit.' You don't want to have that happen. I think it's really important for us to maintain our femininity under all conditions." Some of the women wear makeup in orbit, some don't, but most consider how they look, and will look, in videotapes for the rest of their lives.

In October 2000, it was Melroy's turn. The shuttle *Discovery,* with Melroy as pilot, left the Kennedy Space Center as planned on October 11, and returned on October 24 to her former home, Edwards Air Force Base in California. The seven crewmembers used the robotic arm to attach the main truss and several other large pieces to the International Space Station. In four spacewalks they maneuvered the parts into working position, thus expanding the Space Station so that the next mission could prepare it for the first resident crew.

But before they could begin work, the shuttle's radar failed. For the first time in its almost twenty-year history the shuttle commander and pilot had to rendezvous visually. Melroy was unperturbed. Fifty percent of the simulations she'd practiced were about radar failure so she was prepared to use her eyes and the manual controls. And yet, it was a surprise. The failure had never actually happened in space. She docked successfully, as if she were guiding an airplane, and came away knowing in her gut that piloting a spacecraft was every bit as challenging as flying low through a fog in Panama.

After docking, as she moved with her crewmates into the ready-to-occupy station, Melroy marveled at how easy it was for her to become disoriented. She kept getting lost and losing things because, she discovered, without gravity even the slightest movement changed her orientation. Yet once she got used to it, she imagined all sorts of possibilities. In the few minutes she had to play, she tried zero-gravity ballet and admitted she wanted to explore it further. "I would like to think about how to design a theater appropriate for zero gravity . . . I mean, there are so many technical questions. How do you do entrances and exits? Where do you put the boundary between the audience and the players? Do you need to devise some kind of unusual tether system where there are handholds in strategic locations so that when you danced you could use those to pivot yourself and roll yourself and control yourself? What will you do with your hair?"

A minor detail, and Melroy knows it, but it is an example of how women might anticipate feminine questions of style and taste and expect their male colleagues to understand. Collins, Kilrain and Melroy agree that in contrast to their initial experiences in the military, the astronaut corps is about as gender neutral as any organization can be that operates with a top down command structure.

In 1990 NASA accepted two military women as mission specialists from the Navy and the Air Force in every class, who, like their pilot counterparts, were "detailed" to NASA but remained a member of their particular branch of the service. They brought with them some of the same kind of experiences of being an unwanted and sometimes abused minority. But rather than rage against the military system, they worked to change it, but not too much. They have succeeded in the system as it was and, they are defensive about the ways their own services treat women, even when the treatment is unfair, even cruel, and they are defensive of NASA as well.

By the nineties, American women engineers and scientists knew that NASA would accept a few of them in every class and they knew that their chances would be increased if they applied through the military. Naval engineers like Annapolis graduates Wendy Lawrence and Kay Hire, and Air Force engineers Cady Coleman and Susan Helms, applied as mission specialists. They stayed in the military even as they applied to the astronaut corps, but unlike their pilot friends, they wanted to be members of the team, not the captain.

Kay Hire, unlike her colleague Wendy Lawrence whose father and maternal grandfather are graduates of Annapolis, came to the Academy without any kind of legacy except ambition and a supportive family. The youngest of three daughters in Mobile, Alabama, she knew that her parents—her father is a land surveyor, her mother a draftsman—did not earn enough to send any of them to college. In pursuit of a scholarship Hire discovered the Navy's ROTC program and went for an interview. It was there that one of the interrogators asked if she'd applied to the Naval Academy. Confused, Hire replied, "What's that?"

It was 1976 and none of the military academies was taking women, but there were rumors they were about to. The rumors were correct, and as soon as Annapolis began interviewing, Hire applied, was accepted, and in 1981 she was graduated with degrees in engineering and management.

Too near-sighted to become a pilot, she earned her flight officer wings as a "back seater," which meant that she did airborne oceanographic research, and moved on to instruct first Navy flight students and then Air Force pilots. This gave her a chance to see how each service operated. Out of school and back on duty in the Navy, Hire was the first American woman assigned to a combat crew, the USS *Kitty Hawk.* Then in 1989 she was posted to the Kennedy Space Center at Cape Canaveral, where she certified as a Space Shuttle Test Project Engineer in charge of checking out space suits and the docking system. From there she applied to the astronaut corps and was accepted in 1995, fifteen years after she had stood with her family in their backyard using her dad's surveying transit, in lieu of a telescope, to peer at the Moon and stars.

Her acceptance brought with it a sudden shift in status with her colleagues at the Kennedy Space Center. The announcement of the new astronaut class had come in the afternoon of the Space Center's annual year-end picnic. She held a press conference and then went straight to the gathering where all the people she knew were waiting to congratulate her. "These people that I had known and worked with for many years came up to me at the picnic and all they had were paper plates, and they thrust their paper plate at me and said, 'Can I get your autograph?' I thought they were kidding. I burst out laughing, and they were serious. They were really upset. I got a pen and started signing paper plates. It felt so weird signing paper plates for people that I'd known and worked right beside for several years."

This was her initiation into the recognition and stardom that came with her new job.

Hire left the Cape and moved to Houston, still a commander in the Reserves. Working in Houston she found herself once again with the Air Force, only this time they were astronauts. She still found them different from her naval colleagues, a fact she explains in terms of their histories and in terms of NASA. Naval habits go back to the old sailing days when ships could become separated and each ship's captain had to make all the decisions immediately. He couldn't wait to check with the admiral, because the admiral wasn't there. The much younger Air Force is land-based and, according to Hire, their officers are instructed to defer decisions up to the next level. "We used to have jokes that say, in the Navy, if it's not in the rules, if it doesn't specifically say in the rules that you can't do this, then go do it. Whereas in the Air Force it's like if it doesn't specifically say you can do it, you're not allowed."

What she never said, but is worth considering, is how the astronaut office, with its commanders and pilots, is more Air Force than Navy in how it works. Space law, a new legal specialty that is still evolving, draws an analogy between spaceflight and the sea, where the captain/commander is king. But in spaceflight as we have known it, the Air Force analogy works better because the commander is almost always in voice contact with Mission Control, giving the astronaut office in Houston the last word. This, however, will probably not hold for flights out of Earth orbit.

Hire sees something of both services in the astronaut corps, especially a vestige of Cold War attitudes, and perhaps a sign of military involvement to come. For instance with people visiting from Russia and establishing business connections with NASA, Hire confessed that, "being a military person, a lot of times I'm dealing with international partners [and think] 'Why do we have to share this information with them?'" It is hard breaking old habits, and perhaps even premature.

Catherine Coleman, an Air Force officer, became an astronaut in 1992, not after four years of military education at one of the academies but from the same Air Force ROTC Boston unit that Pamela Melroy had led. Coleman earned a doctorate in polymer science and engineering before she entered the astronaut corps as a mission specialist.

A self-described "military brat," Coleman grew up in South Carolina with an older brother and a younger brother and sister, but when she was twelve her parents divorced and she completed school in northern Virginia. In high school she studied ballet, read a lot and "hung out in a group of about eight, mainly guys, that were in the sort of advanced learner kind of classes. The guys all planned to go to MIT, indeed thought it was the *only* place to go, but [they] assumed that she would not." So she applied, just to show them, with no intention of actually going there. But after her mother made her look at it she decided that MIT was full of people just like herself, people who wanted to enjoy college, with sports and music, as well as pursue their academic work.

Coleman learned more than engineering at MIT. She was introduced to feminism by a friend who was actively involved. At first, Coleman says, she didn't understand why her friend "got excited about these things." But a few years later in graduate school at the University of Massachusetts she told herself, "*This* is what [her friend] was talking about." "This" was the attitude of her young male advisor, who startled her when he confided that there was one guy who was really going to go far, but he was unsure about one of the women. He commented that the woman was "going to have to probably change her image a bit if she wants to make it." Coleman asked what he meant and he told her that the woman in question dressed wrong. Coleman remembered that the woman was "really neat . . . just always dressed in a wild and interesting way, nothing that you would consider too sexy for work or something, whereas this guy that he talked about wore torn-up dungarees with a plaid shirt and hung out at the local bar." When she pointed out that he had not told the fellow that he needed to do anything about his image, her advisor said, "You're telling me I have to think about this a little more, aren't you Coleman?"

Coleman also found herself in a stereotypical female position. She knew that people often put her into an "Oh she's just a pretty girl" category and could not believe that she could be good at whatever she did technically. She had entered MIT in 1978 and in 1981 had heard Sally Ride talk about her life as an astronaut, before Ride had ever flown. Ride's talk convinced Coleman that Ride, too, could have suffered the "just a pretty girl" put-down but it hadn't stopped her. Coleman realized that this was the job she wanted. She would be involved in some educational work, and she would get to fly jets and go into space to explore. It was everything she had dreamed of. Coleman had found her career.

Commissioned a second lieutenant in the Air Force, in 1983 after receiving a B.S. in chemistry, Coleman worked on a doctorate in engineering. When she was almost finished with her dissertation, she reported for duty at Wright-Patterson Air Force Base in Ohio. While there she volunteered to be a test subject at the nearby Armstrong Aeromedical Laboratory, which was doing medical trials for NASA. Once or twice a month she visited their centrifuge for a "workout," in part because she had already applied to the astronaut corps and wanted to know what it felt like to be in orbit. In the course of these workouts she set several endurance records. She recalls that five of the twenty people on the volunteer panel were women, who all basically outperformed the men. She wondered if this was because a different caliber of women volunteered for something as strange as this. Or maybe it was that women were tougher.**

Coleman learned she had been selected as a mission specialist while working in a lab at Wright-Patterson. One of her friends there was a test pilot, "a really neat woman. And yet, here I'm someone who gets picked to be an astronaut." Coleman thinks it may have been because the guys she worked with liked her because "I didn't actually threaten them." But her friend, a pilot, did. Coleman was going to be an astronaut, but not a pilot astronaut. And, Coleman concludes, "I think that being a pilot is one of those last vestiges of macho [culture]."

As an Air Force pilot herself, albeit of small airplanes, Coleman noticed a subtle kind of bias. Men often complained that they couldn't hear women's voices in the cockpit because the frequency was difficult to understand with Air Force radio and communications equipment. But if that's true, she pointed out, it's because the equipment was designed to optimize the frequency of male voices and could easily have been improved. It was certainly true in the sixties; Mary Cleave, one of the two women accepted into the corps in 1980, recalled that when she was a teenage aviator the air traffic controllers had trouble hearing and understanding her high-pitched voice. Yet this kind of incident, Coleman explains, was an Air Force problem. During her time as an astronaut, she reports, nothing like this ever happened. On the contrary, she feels the astronaut corps is an egalitarian place where "people are evaluated for their abilities."

Military women arrived with a variety of experiences. The Air Force pilots came from a different military culture than the Navy pilots, and all the female pilots had different leadership styles from the men. As pilots, female and male, they had separate career trajectories from the mission specialists, which under-

** That suspicion has been a restrained refrain from the earliest days of spaceflight. The idea that women are naturally more suited to space travel keeps coming up, and keeps being dismissed. The most recent voice is William J. Rowe's, a retired physician who published the first of several scientific letters in 2000 suggesting that NASA send only very young women on long-duration spaceflights because young women, during menstruation, slough off excess iron and retain magnesium, both crucial to cardiovascular health. While Rowe's conclusions are eccentric, his estimation that women are the best candidates for space travel concurs with studies that began even before Randolph Lovelace's studies in the 1960s.

lined the inherent inequality in the corps. And while military and civilian mission specialists had a great deal in common, they sometimes differed on the necessity of a chain of command. The military women, both pilots and mission specialists, tended to understand the necessity of running NASA on a military model. They would agree with Eileen Collins, who says, "A chain of command is very important. The military has a chain of command where you report to your boss and he reports to his boss. That's the way things are here also. I think to run a very technical organization and also an operational [one], where you're flying airplanes, flying space Shuttles, you've got to have a structure. People have got to know 'this is your job, these are your limits, this is who you report to.'. . . People's lives could be lost flying airplanes and space shuttles if you don't have a very good strong system in place."

The civilian women who joined the astronaut corps left jobs as physicists, engineers and doctors in the private sector to become civil servants. Some had already worked for NASA and thought they knew what to expect; but as astronauts, as opposed to scientists or engineers, they found themselves working in the quasi-military organization Collins described.

That structure sets out the career ladder clearly. Pilot astronauts fly twice and are then promoted to commander. This ladder distinguishes their career path from that of mission specialists. After their flights, military mission specialists take leadership roles in the astronaut office, or step back into the regular military, receiving promotions in the ordinary way. Both know where their careers lead. Civilian mission specialists, on the other hand, are assigned to an assortment of jobs that are not necessarily linked with anything they did before. Except for physicians, who can make arrangements with hospitals in the Houston area, they are in a career limbo.

Whereas civilian astronauts like Anna Fisher could work part-time when their children were small, military women do not have that option. They have to work full-time. Kilrain, who left NASA in 2003 for a Naval job with more regular hours and closer to her husband explains, "She [Fisher] is a civil servant. I'm in the Navy."[††]

The physicist Kathryn Thornton qualified as an astronaut in 1985 and, in 1993, during the second of four missions, she made two seven-hour spacewalks with her fellow mission specialist Tom Akers—setting a record for length of time for a women doing an EVA—to repair the Hubble Space Telescope. Before she retired in 1996 she had seen the astronaut roster expand,

[††] An exception is Ken "Taco" Cockrell, a civilian astronaut who had retired from the military before joining NASA. He was chief astronaut from 1997 to 1998.

but the changes she witnessed seemed small. The corps was still overwhelmingly staffed by white men. It was still military because the head of the astronaut office, as she pointed out, has never been a civilian. Only a commander is in a position to hold that job and it could be decades before *any* female pilot attains seniority. The social walls Thornton noticed separated pilots and nonpilots and the military from civilians. But like Coleman, she found NASA—for all its nuanced differences—to be a gender-neutral workplace.

Ellen Schulman Baker joined NASA in Thornton's class, having applied to the corps as a flight surgeon at the Johnson Space Center. An exception to the rule that mission specialists do a bit of everything in orbit, Baker has flown three missions in which she conducted only medical or biological experiments. A doctor's daughter, Baker earned her medical degree in 1978 and a master's in public health in 1994. She is comfortable with the hierarchical organization of the astronaut office, which never, she says, inhibited her from speaking her mind. Moreover, "having a chain of command is not necessarily a bad thing." Without this discipline, she points out, there would be 160 astronauts pushing different opinions and jostling for advantage, a working environment that would be inefficient and possibly dangerous.

Ellen Ochoa, an electrical engineer, joined NASA in 1990 along with Eileen Collins. She had earned a doctorate at Stanford and was working at NASA's Ames Research Center when she decided to apply. Ochoa did not decide to major in physics until she was halfway through college and had to choose between two passions—music and science. She does not think that gender affected her career, but she feels that it obviously makes a difference to the kids she talks to at schools, letting them know that she is both Hispanic and female and having an exciting career. But like Thornton, she misses having civilians and women in NASA's management. She observed in 1999, "There may be someone in human resources. But in terms of anybody in my management chain, they're all military still. . . . There are one or two civilians in deputy roles, but never in a direct management all the way up to the top. They're not all active military now, but they have the background."

Shuttles are spaceships, and like ocean-bound ships they are at the mercy of storms, turbulence and dissension within the crew. Space stations, especially stations committed to sheltering a scientific laboratory, may call for a different kind of management. Laboratory culture is not hierarchical and research science is seldom successful with a dictator at the helm. Shannon Lucid, after 188 days on *Mir,* thinks that the International Space Station can do without a military-style command: "You don't need a commander in the sense of a commander of a space Shuttle. . . . I would even change the name.

I'd call him [*sic*] the facilitator. You'd have a person who could make sure that people are working together and producing well."

Lucid knows that NASA has always had a split identity—a civilian agency run as if it were part of the military, a holdover from the Mercury days when every astronaut came through the military. Lucid acknowledges that "shuttle flights are different [from space stations] and there is always the possibility of a last-minute [emergency]." She suggests that "what you need [on the space station] is someone with the ability to act in an emergency, like a fire. 'OK, we need to evacuate.' But the majority of the time you need someone [like a facilitator]."

Lucid, now chief scientist at NASA headquarters, has voiced this suggestion "to everybody who would listen."

Have they listened?

"No," she acknowledges, not yet. "It's very hard to change cultures."

Hard to change cultures by mandate, perhaps, but astronaut culture has already changed from the days of the Mercury Seven, when a handful of male test pilots were processed and molded through the cookie-cutter approach of late-fifties public relations. NASA changed for the first time in 1978, with the enlistment of women as mission specialists. It changed again in 1990 with the admission of female military officers used to functioning in "a man's world." Knowing the strengths and limitations of that world has given these women confidence that they could maintain their femininity, should they choose to, while ushering in changes to an environment they have mastered, and generally—but not unstintingly—admire.

11

AT HOME IN SPACE

ON MARCH 9, 2001, forty-three-year-old Susan Helms unpacked her belongings on *Alpha*, the new International Space Station. She had closed her apartment in Houston and moved everything into storage. The Space Station would be her address, her home for the next five months. Only two years earlier NASA had wound up the twentieth century with a huge send-off for Eileen Collins as she became the first female shuttle commander. Now Helms, the first woman to live on the International Space Station left Cape Canaveral without the public relations extravaganzas that had accompanied other first-woman-to-do-almost-anything-in-space milestones. Women astronauts were no longer news. Her assignment was just another assignment from the astronaut office.

The year 2001 also saw a change of administration in Washington. With the election of George W. Bush to the White House, Sean O'Keefe replaced Dan Goldin—who had set a record for longevity in office—as NASA's administrator. It was also the moment when NASA, with the International Space Station ready for occupancy, focused its attention, at last, on the next step. This step, which policy makers inside and outside of NASA had discussed since Apollo, was sending human beings to explore beyond Earth's orbit. The plans were startlingly ambitious, beginning perhaps with a return to the Moon, or going directly to Mars and beyond. Yet everyone involved acknowledged that these missions couldn't proceed until more was known about how women and men could endure long periods of travel through space. It would not be easy to find out.

O'Keefe, who inherited a gigantic budget deficit, was selected for the job because of his reputation as a budget balancer. Responding to White House expectations, he acted decisively. As a start he halted the construction of the promised new escape vehicle for the Space Station, thereby freezing the Station's crew at three—the number that could fit into a Russian *Soyuz*—instead of the seven that had been originally planned.

This meant ignoring, or reworking, NASA's agreements with its international partners and limiting the ISS modules to those already built and in place. O'Keefe insisted that the halt would not cripple the Station's scientific program. Russia, with the largest slice of entitlement to time and space on the Station, began selling pieces of its share to those who could pay. These included the European Space Agency, which paid Russia to send the seasoned French cosmonaut Claudie Haigneré for a week's scientific research, and Dennis Tito, an American millionaire who paid $20 million because he really wanted to visit the Space Station. Both visitors did some scientific research, maintaining the fiction that the ISS was only about science. NASA issued a statement that the United States would send only professional astronauts to the Station for their part of the time-share.

The international partners agreed that the station was primarily a laboratory, whose scientists' greatest responsibility was perhaps research to prepare and enable human beings to explore deep space, beyond Earth and its Moon. NASA needed to find out if, and how, people could live and work in such a forbidding environment. With the space station now a reality, the psychological and biomedical aspects of adjusting to long-duration flight rose to the top of the agenda. On the station, NASA would gather its own data, which it could add to the data on human responses to zero gravity that the Russians had collected from the *Salyut* missions and *Mir.*

NASA could now focus on the study of human subjects—astronauts—as they responded to extended periods of weightlessness while bathed in a dazzling array of damaging forms of radiation. Not since Apollo, it seemed, had so many people agreed on NASA's mission. In that long-ago era the mission was simple—America had to beat the Soviets to the Moon. America did, and then abandoned the Moon because the space race had been about getting there first, not exploring the Moon scientifically. Lunar science had been an add-on, soon forgotten by most people in the aftermath of the landings. With the International Space Station, a new and promising era was opening in the history of NASA.

The elaborate structure had been on the drawing boards for thirty years and had required a prodigious amount of work on the part of the participating space agencies. Now that it was a reality, it represented the beginning. People would live and work in orbit and soon, or as soon as possible, some would leave to explore other bodies within the Solar System.

The plan for the completed station included seven astronauts in residence. After the budget cut, America's international partners—Japan, Canada, eleven members of the European Space Agency, Brazil and Russia—had to be satisfied with a three-person or even a two-person crew. Each day the work of at least one and a half persons would be required to keep the Station going—which meant that each day only one and a half persons would be available to do scientific research.

That did not faze Susan Helms, who arrived with her colleague Jim Voss in August 2001. Commander Yuri Usachev, the cosmonaut who had been on Shannon Lucid's *Mir* crew, was there to meet them. He had arrived earlier in a *Soyuz.* The three had trained together in Russia and knew each other well. Not surprisingly, Usachev, the commander, immediately offered Helms the privacy of the American laboratory module just as he had offered Lucid her own module on *Mir.* Helms, like Lucid, happily accepted the offer. She realized it was a cultural gesture, the way Russian gentlemen treated women, with a kind of chivalric deference. In Russian society this kind of gesture may have had invisible strings attached that tied the woman to a domestic payback. But on the Station, Usachev could not expect any quid pro quo.

As the Station's commander, he never limited Helms' autonomy. Three days into their visit Usachev remained inside the Station while Helms donned a space suit and stepped outside to install hardware onto the exterior of the laboratory module, a procedure that took almost nine hours and set a new record for a spacewalk by anyone, man or woman.

Nine hours in open space would have fulfilled the wildest expectations of any once space-struck teenager, but Helms had not been that kind of kid. She had never dreamed of a career in space, nor even thought about it until she was an adult, because that kind of career, she explained, just wasn't possible for an American girl born in 1958. She had grown up in Oregon with an Air Force father and a schoolteacher mother, who had worked, she points out, when most mothers didn't. Helms is the oldest of three sisters, all of whom went to college with "the mind-set that there was no point in going to college unless you're going to do something with it." The something she dreamed of was becoming an engineer and, to that end, in 1977, she joined the first class at the Air Force Academy that took women.

At the Academy in Colorado Springs she met people for the first time who looked on the world differently from her Oregon friends. There were about 1,500 cadets in the freshman class that year, including 157 women. Of the 899 who graduated four years later, 98 were women. Helms was sensitive to the shock and hostility she sensed in some of the male students who had suddenly found themselves in an environment that had recently been exclusively male. The Academy had had less than a year between Congress's mandated integration and the influx of women. Some of her male classmates resisted the change; even more just did not know how to handle it. "It was a challenge for everyone, from the highest generals down to the freshmen. You have to realize that the previous three classes to us, all of those male students, expected to go to an all-male school. And then this changed on them right in the middle of their schooling."

Everybody worked hard to muddle through. Most of the women decided to lie low. At first Helms noticed a lot of sexual stereotyping, but as time went on she began to see that people were singled out, not because they were female, but because of what they did as individuals. Obviously proud of the Academy, she added, "by the time we graduated, the place was fully integrated."

With her engineering degree Helms proceeded to Stanford to earn a Ph.D. in aeronautical engineering, and just happened to hear Sally Ride speak after returning from her second flight. It was, Helms realized years later, a standard post-flight presentation, but it convinced her that she, too, wanted to be an astronaut. With her eyes on that same prize, Helms practiced her flying skills while she also worked on the dynamics of weapons separation—preventing the aerodynamic interference that can occur when weapons are released, causing the weapons to return and hit the aircraft that loosed them. After a year and a half in Canada as a flight test engineer, she returned to the Academy to teach aeronautics. There, in 1990, she learned that she had been accepted into the astronaut corps and two years later, in 1993, she flew the first of four technology-oriented shuttle missions.

By 1999 she had flown on four shuttle missions, doing everything from deploying satellites to participating in life science experiments. With a backlog of astronauts still awaiting a first flight, she knew there would be a long wait for a fifth assignment. While she lived in Houston she had followed the construction of the space station with a combination of awe at the enterprise and a feeling that she did not want to go there. She was not the only one. Not much had changed at the Johnson Space Center since *Mir,* when the prospect of learning Russian, living in Russia in the winter, and leaving home for months in orbit won few takers. Astronauts on the International Space Station would also have to devote almost a year preparing for the mission. Helms had little enthusiasm for the prospect, but when she was asked to volunteer, she did. "Just to get back into orbit, I said I would take the job."

Helms expected life on *Alpha* to differ enormously from life on the shuttle, but she did not anticipate how the differences would turn five and a half months in orbit into a remarkably positive and unexpectedly satisfying experience. She points, for one thing, to the matter of space—not "outer space" but the ordinary kind of space that people move around in. Where six or seven shuttle crewmembers live in an area "equivalent to the volume of a Volkswagen bus . . . [the] Station comprises four bus-sized wings that stretch 171 feet." That left plenty of room for the three of them to move around and to be alone when they wanted to be. There was space enough, and also time.

NASA had learned from problems with the three-man *Skylab 3* crew in 1974 that physical work takes a lot longer in zero gravity. Before Mission Control had accepted this truth, the exhausted crew had grown furious with the pressure they felt from ground control in Houston urging them to get work done in an amount of time that may have been reasonable on Earth but that was impossible in zero gravity. The astronauts' frustration grew during the first thirty days of a scheduled eighty-four-day mission, and then they balked. They had had no free time at all since their arrival, not even the planned day of rest every ten days, and they had no other way to convince ground control how difficult it was to perform even the simplest tasks in zero gravity. So they cut communications for twenty-four hours.

When even that did not stop ground control from increasing the pressure for results, the *Skylab* crew enjoyed the first ever space-to-Earth "sensitivity session." After talking out the situation, they rearranged the schedule and eventually did everything that was expected of them—ahead of time. Ten years later, as Soviet space stations orbited the Earth, two-men crews of exhausted cosmonauts managed to work only a few hours a day after months in orbit. They were fighting both fatigue and depression.

NASA and the Russian Space Agency combined their institutional memories to make life on the ISS agreeable. Helms had not been on *Mir* and could only compare the Station to her experience on shuttle missions. In contrast to the shuttle's tight work schedule, on the ISS there was "all the time in the world." There was time to work, time to exercise for two hours a day and time to eat meals together. On a shuttle, work schedules had been so "tightly packed," Helms recalls, that there were many flights when the entire crew came back and said they hadn't had time even to look out the window. A shuttle mission, she explained, is like "a quick business trip," while "life on Station is like a normal workday."

By 2001, space physicians knew the importance of regular exercise for space travelers, and contrasted the condition of *Mir*'s long-duration cosmonauts, who returned weak and unable to walk at first, to Shannon Lucid's success with the boring but efficient treadmill. Between the Russian Space Agency and NASA, space scientists had amassed a lot of data about the way men respond to weightlessness. But they had very little data on women. This would have been a good time to begin collecting data by gender, but there was no plan to study any physiological differences between women and men except, perhaps, what slipped in accidentally. Lucid's strong bones, for example, were noticed, but whether this had to do with menopause, or hormones, or genes, or anything else was left unasked and unanswered. The only conclusion was to encourage every astronaut to use the treadmill.

The ISS crew had a more varied exercise schedule than Lucid had had on *Mir.* Helms, who had danced as a child, developed her own workout in which she bounced back and forth between the wall, the floor and the ceiling, doing "aerial gymnastics and spins and half-turns and all this kind of stuff" that she knew she'd never have another chance to do. That activity was hers alone. A skill that the whole crew perfected was space locomotion. "Your hands end up getting calluses built on them because you are using your hands all the time. You never use your feet [except] to stabilize in place. It's almost like the opposite of what you do on Earth." At times a rookie visitor to *Alpha* got trapped in an open space where, unlike on the shuttle, you can be out of touching distance from a wall. In fact, Helms explained, station crewmembers liked to trap visiting shuttle astronauts, who are eventually wafted to the nearest wall by the air currents from the air-conditioning.

Among the curious but frequent oddities of life in microgravity are efforts to catch escaped globules of orange juice, or sessions talking to each other upside down.

Thinking ahead to a trip to Mars or an asteroid, it is important to prepare for the time when these novelties wear off and you *must* catch the globule of juice because you know if you don't it may gum up the air-conditioning. At this point the crew will face long periods with the same small group of colleagues and very few ways to relieve boredom. On these journeys, for a start, astronauts would be living in confinement for very long periods of time—years on end—without the consoling view of the Earth below.

What would such a crew to Mars look like? Candidates for these slots have changed over time. No one thinks now that astronauts have to be supermen or, as compared with the women selected in 1977, superwomen. In the speculative fiction of the fifties these crews were all male. As recently as 1989, the Soviets, who had always analyzed the psychological makeup of their crews, recommended that any Mars mission should be all male and suggested an even number, perhaps six or eight. Since then there has been a sea change in the thinking behind the composition of this still hypothetical crew. Now NASA and its international partners envision an odd number of men and women, perhaps to avoid coupling, yet all still determinedly heterosexual and multicultural. Any mission lasting several years might be expected to see personal alliances and perhaps shifting heterosexual or homosexual pairings, but NASA has not publicly discussed how to manage these situations. Instead there has been the assumption that whatever jealousies and alliances may arise would not overheat because the crew would be highly trained and professional.

The size of the crew will be limited by the necessity of carrying supplies for several years, including supplies needed for landing and establishing some kind of base. All candidates will have to qualify on psychological grounds different from those used to select astronauts for shuttle missions. Instead of just eliminating candidates who seem unstable—which is the way things worked for a long time—NASA and the other members of the international space community will seek astronauts with a history of stable relationships. They would look for people who have demonstrated sensitivity toward people from different cultures and are able to work together in the unusual intimacy of a spacecraft.

The Russians, far more than the Americans, had paid attention to the dynamics of crew relationships since their first space stations in the late 1970s. They had observed that men who might have begun their mission as friends developed interpersonal problems in orbit. NASA, in "analog studies," that is, studies of people in similar, isolated places, looked for clues to how people would manage on a station in space. These studies, as well as studies conducted by the Russians, show that when personal conflicts erupt in confined groups, the problem is seldom any individual's behavior but the result of the cumulative stress of being stuck together with no respite. One *Mir* crew refused to be in the same room after landing, and are reputed to have avoided being in each other's presence for months after their return.

Helms, Voss and Usachev knew about these studies, about some of the psychological land mines that could mar their mission, and Helms recalls that they "talked about how we were going to make sure that this arranged marriage was going to be extremely successful." They planned to avoid the problems that often developed among scientists in Antarctica during the long winter. In some ways they were at an advantage on the Space Station because each astronaut had a skilled job to perform and there was no social stratification. Another plus for the ISS is the "awe" factor—the wonder of the situation bonds the crew in the knowledge that they are sharing a rare human experience.

There are other key differences between wintering over in Antarctica and living in a spacecraft. On a trip to Mars, the spacecraft would be out of range of rescue a few days after leaving Earth's orbit, the length of days would shift constantly and the whole night sky would look unfamiliar. The awe factor would be extreme, along with the crew's knowledge that they would have to depend only on each other. As electronic mail slowed down, the crew would have to face any emergency that might arise. All the while, the physical work of housekeeping and maintaining the spacecraft would take a lot of time, just as it did on *Skylab.* Most foods would seem bland—astronauts and cosmonauts had remarked on this, and on *Mir* the Americans had noticed how fond the Russians were of the little packets of ketchup and mayonnaise they took for granted at fast-food restaurants. And unless some new form of artificial gravity or a new kind of accelerated engine is developed, the crew will have to exercise for long hours to fight bone and muscle loss.

No one knows how a Mars-bound crew would react psychologically, but NASA knew from shuttle missions that some astronauts behaved strangely even on short missions, and strange behavior would be extremely worrisome on a mission beyond Earth's orbit. Potential problems were noted by Patricia Santy, a psychiatrist who came to the Space Center in 1984 as a flight surgeon and worked with a number of shuttle crews, including *Challenger*'s. She replaced Ellen Baker, the physician who had by then become an astronaut and with whom she worked during her six years at the Space Center.

The hazards to sanity of long-duration spaceflight have teased the imaginations of space enthusiasts since they began discussing the possibility in the early days of *Sputnik.* Coincidentally *Sputnik* appeared at about the same time that American POWs from Korean camps were being released and demonstrating severe psychological problems—a legacy, in part, of prison isolation. At this time both the American and Soviet militaries experimented with sensory deprivation (usually keeping subjects in dark, silent tanks of water at body temperature for long periods of time).

There were sages in both nations who suggested that the veterans' problems were evidence that human beings could not survive in orbit, where isolation from humanity would drive them insane. Their warnings died away after the first men orbited the Earth in the early sixties. The Soviets began their space isolation studies in 1963 at the Soviet (now Russian) Institute for Biomedical Problems in Moscow. In the United States, NASA and the contractors Lockheed and Boeing also ran isolation experiments. During the first eighteen years of spaceflight, these isolation studies, like the astronaut corps, were for men only.

Before *Sputnik* was launched, the first American scientific team arrived in Antarctica in 1957 as part of the activities connected with the International Geophysical Year. Antarctica, a frozen continent, is unique geographically. It is also unique politically, since the 1959 Antarctic Treaty dedicated the continent's entire 5 million square miles to the "freedom of scientific investigation." Subsequently, many countries established research centers there, including the United States, which built several, of which the most ambitious is McMurdo Station, with smaller satellite scientific stations further inland.

In the Antarctic, the long dark winter and low temperatures resemble the hostility of outer space. Throughout the sixties the United States supported a mix of scientists and blue-collar workers "wintering over." They were all male.

In 1969 four American women spent a few weeks in a field camp nearby, but not actually at, the McMurdo Station, and were ridiculed in the press. But by 1974, the momentum of the women's movement had begun to change the nation's scientific and academic culture and two women—a nun and a lay biologist—spent an entire year at McMurdo with no problems. There had to be two women, not one, in those days, and that one of the pair was a nun undoubtedly raised the male comfort level. Five years later, as the first classes of women moved through the military academies and the first women astronauts completed training, a lone female physician spent a year with nineteen men at South Pole Station. Times had already changed. There were no serious repercussions, but she did make an impression. A commander of the Naval Support Force who led several "winter overs" said that the presence of women, first one and then several, changed the tone of the stations. His observation was similar to that of the Russian spokesman at Star City in 1982, who had said that the presence of Svetlana Savitskaia on the *Salyut* was "stabilizing." A woman in residence, the naval commander concluded, reduced fights and rowdy behavior. Perhaps, he suggested, men behave differently when they worry about what women will think of them. Moreover, and perhaps more important, the commander reported that the men were more productive than they had been before.

On the down side, another leader noted, the presence of women added sexual tension. He explained that if a woman established a relationship with one of the men, she usually chose a higher-status Navy man, triggering friction between the white-collar military and blue-collar civilian workers. Not surprisingly, some women complained about unwanted male attention.

The McMurdo crews, isolated as they were, usually numbered around eighteen—about three times the size of the Space Station crew. It is hard to see how, assuming a heterosexual crew, similar scenarios would play out on a space station or in an interplanetary spacecraft.

A hint of the atmosphere on a space station during a long mission comes from some comments of cosmonaut Musa Manarov, who spent a total of a year and a half on two *Mir* missions. Life is difficult enough, he said, without adding women to the crew. He compared a woman on board to a loaded gun: "It could misfire. But if you didn't have the weapon at all, it wouldn't go off." This is not just a Russian perspective. Astronaut Norman Thagard, who also lived on *Mir,* said he "would be in favor of single-sex crews simply because sexual competition in an isolated environment causes serious problems."

If a single-sex crew is better, why not send women? This suggestion has never gained much support, but the possibility has been raised from time to time. The Soviets had, of course, twice planned all-female crews, first in the sixties when Sergei Korolev wanted to see how women would manage, and again in the mideighties when Valentin Glushko again wanted to preempt the United States. Aside from generally weighing less, consuming less and producing less waste, women also seem to quarrel less and are, according to some studies, less easily bored. By the nineties in the United States, most shuttle flights included women, and women had flown on Mir. But the day of the hypothetical all-female crew seemed over until NASA administrator Dan Goldin floated the idea once again in 1999 at a press conference; there he was quickly silenced when the women astronauts rejected the suggestion overwhelmingly.

Sensitive to the press, the women generally agreed that an all-female anything would be called a "stunt" and ridiculed, just as Tereshkova's record-making flight had been in 1963. Ellen Baker, three-time flyer and now working in the life sciences office in Houston, speaks for most of her colleagues when she says: "We feel like we're pretty much equals in the office and are uncomfortable with anything that singles us out for gender alone and not for our technical expertise or experience."

Sending a female crew whose primary mission would be either medical or psychological studies of women astronauts was never on the agenda. One of the few voices defending the idea was Kathie Olsen, NASA's chief scientist between 1999 and 2001. She explained: "I am for an all-female crew for several reasons," the most important of which was developing a set of studies with more than one woman. She liked the idea of having a set of seven individuals they could study at the same time under the same conditions. She acknowledged that "we don't have enough [biomedical] information on men, but we have even less information on women."

This is partly because over the decades NASA has had mixed directives about the importance of science in its overall mission, and of biomedical science in particular. Before the first shuttle flight in 1978, scientific research had received some applause but little real support, with only a few Apollo astronauts looking for rock specimens to bring back. When scientists got a chance to do experiments on the shuttle, it was as likely to be a physics or chemistry experiment as a life science experiment. Since the shuttle began regular flights in 1982, more than thirty missions have carried entire laboratories in the spacious cargo hold, of which a series of twenty-seven experiments were conducted in Spacelab, a collaboration between NASA and the European Space Agency. Spacelab missions continued even after another laboratory, built by a private American company, Spacehab, went into service. Spacehab had two smaller cylinder modules that hooked together or were sent up alone. Here astronauts performed experiments that originated in industries and universities as well as in NASA's own research groups.

Some missions specialized in collecting biomedical information, much of which, like the research of payload specialist Millie Hughes-Fulford, focused on the molecular chemistry of biological problems, especially the calcium loss resulting from demineralization of bones. In space, astronauts on shuttles begin to lose calcium and other minerals within three days; on *Mir* cosmonauts lost, on average, more than 1 percent per month. It seems that most, but not all, people recoup the loss on their return to gravity, but this still leaves a question about the potential fragility of astronauts on a mission to Mars. The symptoms of bone demineralization that resemble rapid aging may be caused by a host of unknown causes in the space environment, and while this phenomenon resembles bone loss in older adults on Earth, the underlying cause may, in fact, be very different. What looks like osteoporosis may demand different treatment for younger spacefarers. However, one of the reasons this particular condition has won so much attention is that a successful treatment for osteoporosis could be offered as another spin-off from space research that would help every American household.

During its first decade NASA routinely monitored the astronauts' hearts and vital signs. Space travel was new and potentially threatening. With the shuttle, as flights became routine, some medical data were collected regularly, but other tests that astronauts complained were uncomfortable were eliminated. Medical data from shuttle missions and from the Americans on *Mir,* as well as from the Russians, who now share much of their biomedical information, are now in a large data base. Some effects of weightlessness are well known: In zero gravity, for instance, body fluids shift upward, with some curious effects. The rise leaves thinner thighs, a result no woman has complained about; but above the waist, where some female astronauts would not bother wearing bras at all, the same lack of gravity makes their nipples stick out in their usual onboard uniform of tee shirts.

One negative result of this fluid shift is Space Adaptation Syndrome, a condition akin to, but different from, motion sickness, which includes nausea and occasional vomiting. After the first women flew in 1983 it seemed that women did not succumb like men, but more data have shown that the discomfort affects men and women at about the same rates. The single exception seems to be pilots, meaning either that pilots adapt more easily to changes of altitude and pressure or, conversely, that people who do not succumb to these problems become pilots. Whatever the reason, two-thirds of all astronauts suffer SAS for one to three days, which represents valuable lost time on a tight schedule.

Another universal reaction is an elongation of the skeleton that adds as much as an inch in height. There are also some data on kidney stones and gallstones and about the difficulty certain medicines have reaching their target in the absence of gravity, as well as the different reactions of medications in male and female astronauts. But because the astronaut corps has 160 male astronauts who have flown versus about 36 women, most of the data are from men.

NASA's life scientists are now as interested in women as they are in men, but the number of women astronauts has never been large enough to get a significant amount of information. The life science office and the public relations office have occasionally been out of sync on this point. While the life science office has bemoaned the shortage of data on the physical condition of astronauts, past and present and especially female, the public relations office had no problem justifying the flight of the elderly John Glenn in 1998. They defended the decision to send him in terms of his contribution to medical information about the response of an old male to weightlessness.

In the years since Glenn's second flight, NASA's life science office refocused its studies to look at individual astronauts, like Glenn, and is thinking about being able to tailor all medical attention in the future to the individual. At the same time they are reconstructing the database they have to get meaningful statistics from all flown astronauts to prepare for those who make the first really long spaceflight, probably to Mars.

The first female candidates for the astronaut corps were screened for conditions like endometriosis. In 1977 women with this history were automatically eliminated because of the fear that zero gravity would result in retrograde menstruation, but that has never been observed, so a history of endometriosis is no longer a barrier. Likewise PMS, premenstrual syndrome, has to be acute enough to interfere with job performance to justify disqualification. The only other exclusively female medical criterion is the amount of radiation women can be safely exposed to, which is less than for men because of the possibility of breast and thyroid cancers.*

Once selected as astronauts, candidates wait to be assigned to a mission, and that can take years. For most of the shuttle's history crew assignments were handled by the director of flight crew operations at the Johnson Space Center, with the approval of the Center's director. These men, and they usually were men, believed they knew each astronaut's temperament, skills and personality. Often they did. But this process did not screen for the personal likes and dislikes of those doing the choosing. Worse, it left the door open to emotional blackmail and favoritism in getting flight assignments.

This seat-of-the-pants system does not seem the best way to choose crews for long-duration flights. Experience on *Mir* taught the Russians what they in turn taught NASA—that a bad crew on a long mission is a preview of crew hell for an even longer mission of space travel, such as a haul over to Mars. Valerii Ryumin, who spent a year on *Mir*, recalls that much of his time was not productive and his productivity decreased continuously until, in his last weeks in orbit, he managed to work only a few hours a day. As for his relationship with his one crewmate, he referred in his diary to an O. Henry story that said, "All the conditions for murder are met if you shut two men in a cabin measuring 18 feet by 20 and leave them together for two months."

The link between psychological and medical criteria is not always neat, which is why psychiatrists like Patricia Santy have M.D. degrees. NASA had always employed flight surgeons like Ellen Baker before they hired Santy. Baker had finished medical school in 1979 and because of her keen interest in aviation, she had found work as a physician at the Johnson Space Center. Once there, she applied to the corps and soon, as an astronaut, she used her medical training as a member of the Institutional Review Board, an internal group at the Center that screens research proposals. In the mideighties she worked with Santy, the first psychiatrist flight surgeon, to examine NASA's approach to astronaut welfare.

Santy approached her job with a different medical perspective from that of Baker and the other physicians. As a psychiatrist she was especially puzzled about the selection process. When she looked for old records, she couldn't find any. Disturbed by the apparent lack of formal procedures, she joined

* Anyone predisposed to developing cancer, including people who have already had cancer and been treated, would probably be disqualified, as might anyone with a genetic potential for developing the disease, because it would be extremely difficult to treat in the environment of a spaceship.

Baker, who was also concerned, in 1988 on a Committee on Psychiatric and Psychological Selection. This was the first working group of its kind in the history of the Johnson Space Center. Baker and Santy decided to ignore the old informal rules and instead isolate a set of psychological traits they believed might be used in selecting future astronauts. The committee presented their conclusions to NASA in 1989.

Looking ahead to long-duration missions, the working group suggested that the space agency needed, at last, to *select in,* to look for special psychological strengths. The ideal astronauts, it seemed to them, should be confident without being arrogant, and have a sense of humor and a generous spirit when dealing with stress—a description that should also apply to all bosses and spouses, and a combination that may well be in short supply.

In an effort to put her research in historical context, Santy tried to review NASA's selection procedures from the beginning, and discovered that there were no records from the Mercury astronauts through the entire shuttle era. Disturbed by this apparent vacuum, she addressed an outside review committee that included Chris Kraft, who had worked with the Mercury Seven before he headed the Johnson Space Center from 1972 until his retirement in 1982. Kraft personified NASA's dormant old boy attitude. To Santy's chagrin he "became agitated and interrupted" after she had shown only a few slides and said to her: "Young lady, you are a dangerous person and are out to destroy NASA! I will not permit that to happen."

When she explained that she had simply been looking for documentation, he replied, "You never will [find it] because it's in here [pointing to his head], and it's going to stay there so that people like you can't use it against NASA!"

Santy described her experiences in the book *Choosing the Right Stuff,* where she records that, Kraft's withering contempt notwithstanding, the Working Group persevered. But it soon faced a new obstacle. NASA ordered them to include homosexuality as a *psychiatrically* disqualifying condition even though the Group pointed out that homosexuality per se is not a psychiatric disorder and they did not intend to disqualify anyone because of sexual orientation. If NASA insisted, then it would be a management decision, Santy said, not a medical/psychiatric one.[†] NASA insisted, probably because it was concerned with its image or feared homosexual activity in space. NASA's social center of gravity was exceedingly conservative and at this time gay people had not been

[†] Patricia Santy says that Harry Holloway (who later became head of Space Life Sciences at headquarters) commented that he wondered if management realized that it was likely that over all the years they HAD (albeit inadvertently) selected homosexual individuals as astronauts. Management was aghast at the suggestion.

accepted by many American institutions, including the armed forces. The Working Group complied, rationalizing their acquiescence by insisting that the candidates would also have to fall short on another count.

By the millennium, after the Russian space station *Mir* had descended ablaze after eleven years of orbiting the Earth, the International Space Station was about to replace it. A new era was beginning and some of the old ways of selecting and training astronauts were about to change.

George Abbey, who had worked with astronauts since Apollo, was still the director of the Johnson Space Center as the International Space Station came together. Abbey had sat on the astronaut selection committee since 1978 and had headed the Center from 1995 through February 2001. He had a record to be proud of. He had selected the first women astronauts and added women to every new class or assigned them to shuttle crews. He had never objected to the formality of selecting out, eliminating astronaut candidates who had histories of mental problems, but he did not consider using psychological tests to help *select in*—that is, to fit together harmonious crews. Often a committee of one, he could have pointed to his own head, the way his predecessor Kraft had done a decade earlier, when asked about the criteria he used for selection. He knew what was needed. His process has been called "capricious," as "at times it seemed that people who actively disliked each other were assigned to some crews." A controversial figure, Abbey, who retired in 2002, is remembered as either a Machiavellian or a benevolent despot.

Most of the astronauts who opposed him had their careers cut short. Currying favor with him obsessed members of the astronaut corps from the beginning. Drew Gaffney, a physician, flew as a payload specialist on a Spacelab mission after working with the Life Sciences Division at JSC for two years. His outsider status freed him to comment: "For astronauts the first rule is, 'If you suck up, you go up.'" Following *Challenger,* Gaffney brought the issue to friends in Congress, and after NASA polled about 20 percent of the workforce in Houston asking them to evaluate their bosses, Abbey's astronaut office came in last. Yet it is said that Abbey had institutional staying power akin to that of J. Edgar Hoover. In the end, it took a total change of administration, and advancing age, to topple him.

However, he *was* benevolent to women. Not all the female astronauts liked him, but he was an advocate of women as astronauts. He had favorites, of course. For whatever reason, he apparently selected Sally Ride to fly first, and then assigned the other women to shuttle flights, first one and then two and three women to a crew. On his watch the presence of women in space became commonplace.

Before the International Space Station, NASA did not need a subgroup of astronauts selected for their ability to work for months at a time in a hermetically sealed, mini-environment in zero gravity. NASA knew from the experiences of its own astronauts on *Mir* that some people were better at this than others. It was becoming apparent that certain personality traits would be needed on the Station and in the crews that would one day make the estimated three-year round-trip journey to Mars.

Susan Helms described how she knew about the McMurdo isolation studies and how they predicted low points in the relationships among crewmembers. She knew about the crisis on *Salyut 6* in 1977, when cosmonaut Georgi Grechko was not told that his father had died during his three-month mission for fear of an emotional response that would endanger the flight. They also knew that on a later *Mir* mission another commander was told about his mother's death, and was able to deal with it aided by his crew. Helms, Voss and Usachev used that experience as a starting point and planned what they would do in case of a death in their families. As it happened, both Americans suffered close personal losses, which they handled well, according to Helms, because they had discussed what to do in advance.

Especially revealing is that they did *not* experience what the analog studies had predicted. Contrary to the people wintering over in Antarctica, who eventually tended to distance themselves from each other, they did not become depressed halfway through their mission. And at the end of the mission, when the studies predicted they should have been eager to return home, Helms regretted leaving. She had grown to love living in orbit and knew that she would probably never again marvel at the awe-inspiring view and the gentle peace of space. Laughing, she added that she would even miss the meals.

"We all ate a lot. Every one of us ate a lot!" And best of all, "I lost twenty pounds. I don't know how, because I certainly was eating my heart out, just eating, eating, eating!" The men lost weight too. "It's typical of long-duration flight," she explained. In contrast, women in bed-rest studies, which are still being conducted, almost always gain weight. Space is unique. Its environment cannot be replicated on Earth.

Yet these limitations notwithstanding, studies still in progress are trying to understand what can be done to enable people to thrive in a society of two or three for extended periods of time. The earliest studies were done in 1961 at the Institute for Biomedical Problems in Moscow and were planned by Korolev and Keldysh, the men most responsible for the well-being of the first cosmonauts. In 1963, the Institute moved to its current site, where the staff, some still there from the beginning, recall testing Tereshkova and Gagarin.

The Soviets and then the Russians were keenly aware of the potential for friction in the alien atmosphere of a space station. After *Mir,* and in anticipation of the International Space Station, the Moscow Institute decided to explore the behavior of a mixed crew—a non-Russian and a woman—in extended isolation.

Late in August 1999, the two Russian men who had begun living in an eerie kind of isolation in the Institute four months earlier awaited the arrival of two women. In the decades since 1961 all previous Russian isolation experiments had used only males. The psychologists wanted to see if the added ingredient of women made a difference to the psychological balance of the "crew." Almost from the start, the experiment went awry. One of the two women volunteers failed the physical, leaving just Judith Lapierre, a thirty-year-old Canadian psychologist, to join the two Russian men. The Russians replaced her female partner with a male Japanese volunteer. He and Lapierre would live in a different part of the isolation area, in separate rooms, with a door between them and the Russians.

The four seemed to be getting along until New Years Eve, when, to celebrate the millennium, someone provided champagne. The four were talking together, and then suddenly one of the Russians grabbed an unwilling Lapierre and kissed her. It was not a friendly peck, she explained to the press at the end of March when the experiment was over, but an open-mouthed sexual advance. Lapierre complained at once to the authorities at the Institute, but no one helped her. A fight broke out between the Russians, and the Japanese volunteer cowered. The Russian kissed her a second time, whereupon the outraged Japanese, along with Lapierre, left the room and locked the door behind them. The Japanese volunteer soon quit the experiment, ruining the plan. Lapierre stuck it out, but when she emerged in March she told a press conference about the New Year's Eve events. The incident had grown in her mind to the equivalent of rape.

The experiment, a success in its failure, revealed an enormous difference in national sensibilities. Lapierre believes, along with many Western women, that no woman should ever have to put up with unwelcome sexual advances. The Russians conducting the experiment tried to explain to her that the rowdiness of their volunteer was a sorry misstep by a man who had been isolated for four months before she arrived. And besides, it was only a kiss.

It was clear that there are potential hazards when a situation calls for collaboration between people from different cultures, especially in heterosexual crews. Lapierre's experience is an example of the kind of behavioral problem, compounded by cultural differences, that NASA has ignored for a long time.

Indeed, for a time NASA managed to avoid considering the whole area of sexuality and sexual reproduction. When the questions about sex finally seemed unavoidable, NASA made the curious decision to collect sex data on fish and rodents instead of people. Human sexuality was a no-fly zone except for NASA's unofficial stand against homosexuality. This was not because NASA's scientists and managers were naive or prudish, but, according to an apologist, because they were afraid that the mention of sex in space would send the news media into a frenzy of distortion, sensationalism and trivialization that they, apparently, had no way of counteracting.

NASA's administrators have been hamstrung from the start by a puritanical strain in the American public. It surfaced, of course, in 1960 with the FLATS when female aviators asked to be considered for the astronaut corps. It flared again in 1972 when the *Pioneer 10* spacecraft, the robotic tourist of the Solar System, was launched. The craft carried a plaque so that if one day another intelligence found it, they would have an idea of who had sent it and from where. The plaque included a map of our Solar System, and line drawings of a naked man and woman. This latter picture appeared in newspapers, triggering outraged letters asking why there were pictures of naked people, images bound to offend extraterrestrials. NASA compromised with its critics and, instead of adding fig leaves as some suggested, removed the genitalia. At the same time the public relations office fended off comments by Charles Berry, one of the first flight surgeons for the astronauts, who had suggested, perhaps as a joke, that astronauts on long-duration flights might need a sexual outlet. He had recommended including a woman, albeit a woman scientist, who could perform what can only be called "double duty." The public relations office continued to describe spaceflight as a bachelor party.

When Santy was working in Houston in 1985, a researcher at NASA's Ames Research Center commented that healthy astronauts might want, or even need, sexual partners. She also wrote that NASA was considering private space in its design for the planned Space Station. The press described this arrangement as "quarters for intimate behavior." Copies of outraged letters to Senators Charles Grassley and Robert Kasten reached NASA's assistant administrator for legislative affairs, who quickly backpedaled and explained that the Ames Research Center was simply studying ways to make life as normal as possible in space. The researcher retreated and the public relations office at Ames closed the discussion with a statement that NASA had not spent a dime studying the effects of space on reproduction in humans or animals.

While the people at Ames investigated comfortable designs for a space station, the medical staff at the Johnson Space Center asked for help from the

National Academy of Sciences' National Research Council. Since its beginning, NASA has had contracts with the National Academy of Sciences, which serves as an advisory agency. Within the Academy there are three branches, one of which, the National Research Council, organizes panels to investigate the sciences and engineering. The medical staff wanted them to assess research on the medical and psychological impact of living in microgravity. In its 1988 report, one panel member, Lynn Wiley, a professor of obstetrics and gynecology at the University of California at Davis, noted that there had never been any effort to get sperm samples from the astronauts.[‡] She summed up the situation as she saw it, saying that "sex and semen are just not topics you talk about with those guys. They shut up like clams, so we have no data."

In 1989, R. J. Levin, a British physician, published an article on the possible impact of space travel on sexual physiology. He, too, observed that "it is remarkable that there is no study as to whether human sexual activity, or its loss, can influence the adverse effects of such environments." He listed a host of possible effects of sexual deprivation, including "boredom, listlessness, sleep disturbances, fatigue, impaired cognition, irritability, hostility, depression, and deterioration of personality."

American society, never homogeneous in its attitudes toward sex, was divided more than ever. The physicians and psychologists who wanted information about physiological effects of living in zero gravity kept asking for information about sex. Unlike the public relations office, they did not have to face the growing constituency of Americans who would not countenance any government research that might lead to abortion, birth control or sexual activity outside of traditional marriage. NASA ignored the issue, perhaps in the hope that it would disappear.

Sexual research may have been kept out of NASA's laboratories, but no one could remove sex from the buzz at its space centers where NASA employees and camp followers mused about the "200-mile-high club," a quasi-mythical organization whose members have experienced sexual intercourse in space. The many candidates included Jan Davis and Mark Lee, who were secretly married right before they flew together in 1992. The club's possible members, however, never include cosmonauts—as if sex were an American proclivity—and the gossip never included the possibility of homosexual couplings. When astronaut Janice Voss heard some of these tales on a spring evening in 1999, she gave a hoot. With a scant eight inches between "hammocks" on shuttles, if two people got together she did not see how they could

[‡] Patricia Santy notes that several of the male astronauts confided in her that they had "tested" ejaculation and there didn't seem to be a problem in zero gravity.

have any privacy. Pamela Melroy concurs: "I can't think of a way that you could do that without every single other person knowing exactly what was going on." Besides, she couldn't see it happening in the kind of crew she knew. Recalling her experiences during the Gulf War as well as her first shuttle mission, she described a special dynamic in a crew working intensely together that forestalls romance. "I'm not saying that there aren't people who have fallen in love. But typically, it ends up falling into what I call a family dynamic." Of course, families experience incest, and a spacecraft in a multi-year mission would be different from a squadron in battle or a crew visiting the Space Station for six months.**

While NASA seems to have avoided studying human sexual behavior in orbit, some people have forged ahead. In 1992 an otherwise unknown Sunday school teacher in North Easton, Massachusetts, patented a device made of belts and loops that allowed "one partner to exercise control of the movements of the hips of the other partner during the act of sexual intercourse" in space. She may have remembered the lyrics of a 1985 recording of Dana Gallagher's "Zero-g Sex":

> Making love in a zero-g environment
> May not produce the satisfaction we project.
> The proposed erotic possibilities
> Deny Newton's laws of motion
> And how they will affect
> Our romantic inclinations
> In space stations.
> Momentum changes in proportion to the force
> Acting on a resting body
> And making it move.
> And every action is opposed
> By an equal reaction
> Of the same magnitude
> In the opposite direction
> Without exception.
> . . .
> Adrift in compartments three-dimensional
> Pelvic push may quickly find

** Patricia Santy disagrees: "As for privacy—well, of course everyone would know what's going on, but there is no reason why a dedicated couple couldn't have the flight deck to themselves for a while, while the rest of the crew stayed below!"

Lovers parted and marooned
Or rebounding off the walls—
Ricochet free-fall.

One of the first people to muse about having sex in orbit was Michael Collins, an Apollo astronaut, who wrote about the possibilities of weightless sex in *Lift Off*: "All float with the same delicacy, aerial ballet dancers all. And lovemaking! . . . having no gravity to crush bodies together offers exquisite possibilities. A tiny pull here, a gentle touch there. A space *Kamasutra* remains for some lucky couple to write . . . lovers in an orbiting hotel could explore their minds and bodies in ways simply not possible here on earth."

Elena Kondakova had no such fantasy. She commented that while she would love to fly with her husband, Valerii Ryumin, for "making love, it's much better to do it on the ground."

That may in fact be true, but the possibility of sexual activity in space, if only for the novelty, appears over and over in just those places that NASA spokesmen fear most. For instance, a British tabloid in 2001 reported—apparently erroneously—that crewmembers on the space station had contraceptives tucked into their medical kits. Susan Helms could not speak for the others, but she knew there was nothing in her kit.

The adventure of space travel began just as birth control pills hit the market and the sexual revolution began, but NASA behaved as if the fifties had never ended. While part of American society demanded the inclusion of women everywhere, including the astronaut corps, another part found the idea offensive. NASA needed the support of both constituencies, so it accepted women into the astronaut corps and avoided all suggestion that sex of any kind was practiced by any of them.

The space center coasted along accepting the posture most of the civilian women had adopted. They were astronauts, not *women* astronauts, and that meant that there were no differences between women and men.

In taking this position, though, they were falling out of step with society. In the early nineties, women began to demand that equal attention be paid to women's health problems, including AIDS, breast cancer and heart disease. In response, the National Institutes of Health announced the Women's Health Initiative, a fourteen-year study to investigate almost every possible influence on the health of post-menopausal women. Then in 1992, NIH directed that clinical trials of new drugs and regulated products include women; until then men had been used almost exclusively. American medicine was finally responding to women's issues.

All the while, NASA's biomedical staff postponed studying sex and sexuality in space. The shuttle was still America's only extraterrestrial destination and NASA deferred the matter of exploring sex in orbit, calling it "premature." This was a curious head-in-the-sand position in the eighties and nineties. For as NASA dodged any sexual agenda, gay liberation and new methods of assisted reproduction were changing the nature of families and certainly the makeup of PTAs and workplaces across the country. As families around the world followed televised accounts of the sexual behavior of nonhuman animals, especially chimpanzees and gorillas, NASA quietly looked into the reproductive lives of the fish and rodents they took into orbit in the cargo holds of the shuttles.

Until nearly the turn of the twenty-first century NASA focused on biomedical rather than psychological problems. But even here the problems they addressed seemed narrow, and in 1987 their restricted scope of inquiry was noted by the Space Science Board of the National Research Council, which had put together a panel of experts to examine NASA's biological research. The panel listed questions that NASA might consider answering in the next decade and concluded with queries about the lack of research on animal models, and the lack of information about the effects of zero gravity on women.

The report asked for information about menstruation and ovulation, and in nonhuman animals about fertilization and the development of pre-embryonic tissue. One suggestion, for instance, was that in the future NASA collect data on the basal body temperatures of female astronauts for at least three months pre-launch. And for the first time they asked NASA to compile complete reproductive histories for all astronauts that would continue for several years post-flight in an effort to determine the effects of long-duration space travel on birth defects and pregnancies. Careful not to single out women, the Board also suggested collecting pre-flight, in-flight and post-flight sperm samples to determine if space triggered any anomalies there.

Instead, NASA's life sciences scientists continued to address unisex problems, or, just as significant, NASA treated most medical problems as if they were unisex in nature. The following year, 1988, a panel of space biologists from the National Academy of Sciences' National Research Council noted that there was *still* a lot to learn about the effects of stress, microgravity and radiation on astronauts.

The Academy and NASA seem to have enjoyed a history of respectful interaction, with NASA asking the Academy for advice and the Academy duly organizing a group that offered suggestions. It is unclear if any of the

Academy's suggestions or those from its own Working Group in 1989 were heeded at the time.

NASA avoided exploring medical gender issues while Congress funded the Women's Health Initiative to investigate the medical problems of menopause. Eager to catch up with the changing attitudes toward women, Bonnie Dunbar, then special assistant in the Office of Life and Microgravity Sciences and Applications, explained in 1994 that NASA's medical data in the last decade had resulted in more than 7,000 published papers, and that "a large percentage of that research is applicable to women's health issues." She was referring, in part, to the study of bone demineralization. NASA's public relations team had extrapolated these studies of bone loss among healthy young people in zero gravity to the problems afflicting older people on Earth. She suggested that NASA's research might lead to an explanation of, and perhaps treatment for, osteoporosis.

It would have made sense at this time for NASA to include hormone studies in its research of astronaut health, but they did not. The researchers explained that there were too few women who had been in space—then about 36, as compared with over 160 men—to get useful data. They did not mention that a major obstacle to gathering information about women's health has been the reluctance of women astronauts to participate in women-centered experiments. NASA depends on the voluntary collaboration of astronauts for all biomedical studies. This really means the collaboration of female mission specialists because commanders and pilots cannot volunteer for any test that might diminish their abilities to handle the craft in case of emergency.

Many astronauts have not complied. Like all Americans, astronauts are protected by human subject regulations from having to participate in an experiment they don't choose to join. As things stand, NASA cannot select a crew on the grounds of their willingness to be test subjects, nor can they force an astronaut to participate who has agreed to do so and then changes her mind.

Within the astronaut corps there is hostility to the idea of being a subject. Kathryn Sullivan, from the class of 1978, refused to be used as a "human guinea pig." Although Ellen Baker sensed that a lot of this attitude had apparently eroded before she arrived at NASA, and believes that her fellow astronauts now participate between 80 and 90 percent of the time, she defends their right to refuse. David Williams, the astronaut who headed the life sciences division from 1999 through 2002, says that he never had a problem with compliance.

However, there remains the unwilling 10 to 20 percent cited by Baker. Susan Helms recalls a shuttle mission in 1996 when four people were needed as subjects for medication experiments. The experiments called for either two men and two women, or four astronauts of the same sex. Since she was the only female in the crew, she didn't participate. This brings up the folly of running an experiment that could have contrasted male and female responses to a medication without bothering to send it with a crew that included an adequate number of women mission specialists. As late as 1996, it appears, gathering comparative medical data about women and men was an afterthought to NASA's medical curiosity.

In fact, gathering data on astronauts at all was a problem, according to Santy. She recalls trying to get medical data to see if the astronauts in her charge were fit to fly, only to learn that a new regulation prevented her from access to even that apparently crucial information. However they sweetened the directive, the fact remains that if a study proposed to measure the response of the heart in space, it would necessitate a pre-flight examination and the astronaut to be tested might reasonably opt out for fear of being grounded.

That, in sum, is the top fear among all astronauts. After a year or two training for a mission and even longer awaiting an assignment—why risk it? Ellen Baker estimates that she spent thirty days in space during her first fifteen years as an astronaut—two days for every year served. It is easy to see why an astronaut would avoid risking those days to take a medical exam in order to participate in an experiment to provide medical data that might disqualify him or her altogether. But Baker defends NASA, quietly insisting that the relationship between doctors and crewmembers is much better now than it was thirty years ago. Then the testing was "what some people might think unreasonable," whereas now, she says, they just get the standard physical.

The completion of the Space Station triggered a shift in the medical agenda. Although still concerned with addressing Earth-bound illnesses, NASA is now focusing on medical problems that result from living in zero gravity, and ways to prevent or treat them. Americans are thinking about sending people to Mars or back to the Moon and it is suddenly crucial to discover how women respond to the peculiar environment of space, how women differ from men, and how women and men can work productively as a team in a spacecraft.

Susan Helms volunteered along with Voss and Usachev to supply medical data, part of which came from participating in a spinal-chord excitability experiment in which electrodes were strapped to their calf muscles. Helms explained: "The point of it was to see how well the nervous system responded pre-flight, in-flight, and then recovery post-flight, as far as response to involuntary contraction from electrical stimulation of the calf muscle." She volunteered because she knew that she could contribute to helping other astronauts who are already planning long-duration missions.

By 1999 NASA faced head-on the changes that the Space Station had brought to space exploration and once again asked the National Academy of Sciences—this time its Institute of Medicine—to help them "create a new vision" of health care outside Earth orbit. The Institute responded at the end of 2001 with the publication of *Safe Passage,* a detailed report on what is known, and what needs to be known, in order to send people safely on long voyages of exploration. The report makes some striking assumptions that its authors don't seem to realize are striking. They assert, for example, that humanity is on the verge of exploring deep space—which may or may not be true—since the political and economic climate for funding and organizing such a venture is always volatile. They also take as given the decision that the spacecraft involved will carry both women and men, but even that is not certain.

Assuming, however, that women continue to become astronauts at an increased rate, the report calls for a plan that assures this generation of astronauts that they will not suffer physical or psychological damage. To this end, *Safe Passage* documents what is known about the effects of microgravity on every area of human health from the muscular-skeletal to the circulatory-pulmonary and on to the alimentary, nervous and reproductive systems.

The report points out that in spite of having flown approximately 400 people since 1962, "a paucity of useful clinical data has been collected and analyzed." While complimenting NASA on its analysis of Space Adaptation Syndrome and muscular-skeletal problems, it faults the virtual absence of research into the differences between female and male astronauts.

The medical black hole about the responses of women and the whole area of human sexuality is partly a result of the astronauts' distrust of doctors. Complicating this already tense relationship with the people who can ground any astronaut at any time are the women themselves. They do not want to be tested for evidence that their biology is different in any major way from that of their male colleagues.

Yet a 1999 study of women astronaut health highlights differences that ought to be examined. One major difference they cannot ignore is the possibility of infertility, but it is hard to pin this entirely on zero gravity. It is rather a problem they share with many professional women who postpone childbearing. However, the fact cannot be ignored that most other professional women aren't flying around in zero gravity under intense bombardment by radiation. The stresses of space combine with these other statistics: The average female astronaut enters the corps when she is thirty-two, and has no children. What with training and awaiting assignments, most do not get around to starting a family until they are forty. Then they face possible infertility. Like other professional women who have postponed childbirth, astronauts use a lot of assisted reproductive technologies. By 2002, ten women who had flown in space had babies at an average maternal age of forty. Some had used birth control pills, others had not. Age seems to be the problem, and NASA is now considering offering embryo freezing and even egg extraction and freezing so that women who have deferred maternity can still have children.

For the foreseeable future NASA will not fly pregnant astronauts and assumes that pregnancies will not occur in orbit. Most of the women applaud this policy. What evidence is available suggests that a developing fetus in zero gravity would suffer grave defects. Reproductive matters aside, most female astronauts believe that differences between individuals are far greater than any broad-stroke differences between women and men and they reject tests for difference as both mischievous and unnecessary.

An undercurrent of fear runs beneath these statements. What if differences are found, and then, as in the past, the differences are held against them? Having spent most of their professional lives insisting that gender is irrelevant to performing their jobs as professional astronauts, women astronauts understandably find any effort to separate women from men threatening. Perhaps this is why, as the Institute of Medicine concludes, a major problem of space medicine research is the small number of astronauts who volunteer as research participants.

When Sally Ride flew in 1983, physicians were discussing "retrograde menstrual flow," the backing up of blood into the abdominal cavity. When it didn't happen, there was no systematic follow-up. It is now, again, on the research agenda, but it is only there because the issue has never been thoroughly explored. Likewise, a study by former astronaut Rhea Seddon, now at Vanderbilt University Medical School, notes that there are no studies on hormone secretion and the effect of space travel on ovulatory function. Neither are there studies on the impact of stress and excessive exercise on decreased estrogen levels and decreased bone mineral density.

NASA also has no data about the probably different reactions of women and men to pharmaceuticals, radiation and psychosocial adaptation. The National Institute of Medicine recommends that Seddon's suggestions, as well as suggestions from those earlier oversight groups in 1987, 1988 and again in 1999 which studied how women responded to every aspect of long-duration flight, ought finally be carried out.

Some of these recommendations have finally been incorporated into the life science program. But most cannot succeed without changing the human subject regulations. These include a limit to medical confidentiality, a limit already used by the Department of Defense that says information can be used to aid in preventive and communicative disease control programs and the conduct of research. The National Institute of Medicine notes what most people understand—that astronauts are different from other Americans. They alone have the rare privilege of experiencing space travel and they are the *only* people whose medical tests can reveal the effects of long duration in microgravity on human health. They alone can show us what has to be done to protect future spacefarers—astronauts, tourists and even salespeople—from the hazards of the still mysterious and alien environment of space. Moreover, the Institute reminds us, they will not be the first Americans to relinquish their privacy. Other exceptions to human subject protection are laboratory workers who use dangerous substances, which is a good analogy because the Space Station is, in fact, a laboratory in a precarious environment. It is hard not to agree with these conclusions.

Yet it remains a difficult issue for the astronauts, who largely disagree. Helms did not participate in a life science experiment on her 1996 shuttle mission, although, with everyone else in the crew, she had signed an informed consent form before crew selection and believes the research was important. But during that flight she changed her mind because the experiment looked painful.

The experiments that most astronauts reject, Helms explained in her defense, are "in shuttle language called a Detailed Science Objective, DSO." She contends that they are not peer-reviewed by outside scientists but originate instead at the Space Center and go through a different approval process that doesn't have "the operational safety" of what she calls "legitimate experiments." Patricia Santy, now at the University of Michigan, suggests that Helms expresses the astronauts' fear that the DSOs will simply reveal too much about themselves. David Williams disagrees with Helms, holding that the internally generated experiments are both safe and legitimate.

Participation aside, even if astronauts do volunteer, they have the right to keep their records private and prevent whatever data generated from becoming part of the data base. The Institute of Medicine lists exceptions to this tradition of confidentiality. These include information from people in the military, in prisons and in the CIA, where medical information is released in order to control communicable diseases or monitor workers' health.

Any spacecraft can be considered an area where occupational health issues exist that threaten a safe workplace. For instance, in 2000 a radiologist at the Johnson Space Center concluded that astronauts who had flown at high inclinations (more than 250 nautical miles above the Earth) developed cataracts a decade ahead of most people.

Astronauts traveling high above the Earth are like the workers in the early nuclear power plants. There are unknown risks in long-duration missions and the only way to create a safe workplace for future astronauts is to know the medical histories of those traveling in space today. Women will play a crucial role in this risky but necessary step to enable humanity to soar beyond sight of Earth and into the unknown.

EPILOGUE

ON JANUARY 16, 2003, while the International Space Station circled the globe with a bare-bones crew and scaled-down research program, NASA launched its oldest war-horse, the shuttle *Columbia.* NASA had declared scientific research to be its primary purpose, and *Columbia,* STS–107, was a mission devoted to science.

Columbia carried seven astronauts on the flight deck and the Spacehab laboratory in its cargo bay. Mission specialists in the crew would perform eighty-eight scientific experiments during sixteen days in orbit. It was very ambitious, and would probably be the last all-science shuttle mission.

Columbia's crew, looking a lot like *Challenger's*, included two female and five male astronauts. They were also as socially diverse, including a Jewish astronaut, Colonel Ilan Ramon, a payload specialist from Israel; an African-American, Michael Anderson, a lieutenant colonel in the Air Force; and an Asian, a naturalized Indian-American woman, Kalpana Chawla. But here the similarities ended.

Columbia had risen undeterred into orbit. A video of the launch shows a piece of debris, most likely insulating foam that had fallen from the external fuel tank, striking the left wing just eighty-two seconds into flight. At first glance, it looked as insignificant as a moth striking a windshield. Engineers at Boeing, the subcontractor responsible, figured that the wing had not been impaired. Meanwhile, unaware of any problem, mission specialists Laurel Salton Clark, David M. Brown and William McCool, all of them rookie astronauts, enjoyed the incomparable sensation of floating in microgravity. The old hands, Commander Rick D. Husband, Anderson and Chawla (who liked to be called KC), found their space legs rapidly and got to work.

Clark and Chawla had different scientific, national and religious backgrounds. KC grew up in Karmal, about seventy miles north of Delhi. Her family were Sikhs, and very religious. Although KC's older sister had studied medicine, their father had been especially protective of his younger daughter and reluctant to let her travel to the nearest large city where she could study aeronautical engineering. Yet she gave him credit for her passion, for he had taken her on a ride in an airplane when she was very small. That experience left her determined to learn to fly, and eventually to design airplanes. Ultimately her father let her study engineering at Punjab Engineering College.

With degree in hand, she came to the United States and in short order met and married Jean-Pierre Harrison, her flight instructor. They moved to Boulder, Colorado, where KC earned a doctorate in aeronautical engineering at the University of Colorado and began thinking about becoming an astronaut. Having married an American and become a citizen, she found an engineering job at NASA's Ames Research Center. When she noticed a bulletin announcing a new class of astronauts, she decided to apply, figuring, "You're not going to win a lottery until you buy a ticket."

She won the lottery, of course, and she never doubted her good fortune. At NASA, she had a rocky start when the robotic arm she operated on her first mission lost control of a satellite, and she was blamed for closing the arm too quickly. Eventually the entire crew was cited for a chain of errors and she was absolved of responsibility for the mishap. Nonetheless, she felt that she had to work even harder.

Away from NASA she loved hiking and camping and described the view of Earth looking down from space as "the good campsite below us, because there is water, there is air and this is the only campsite we know." She wanted to share her vision with friends in India and brought girls from her old school to visit her at NASA to inspire them to follow their dreams as she had hers.

Laurel Salton Clark, although a rookie in space, was an experienced underwater physician. Growing up in the Midwest, away from the ocean, she had wanted to be a pediatrician and joined the Navy to pay for medical school. She had planned to leave the Navy after fulfilling her obligations, but discovered that she loved the sea and practicing medicine in a submarine. So she stayed in the Navy. When she was ready to move on, she applied to NASA as a mission specialist. She liked students and on one school visit went to her college roommate's classroom at her middle school in Milwaukee. She brought along a stack of books five feet high, all from the courses that astronauts take, to show how astronauts never stop studying. She was married to another physician and when she was not at work she spent time with him and their eight-year-old son. She was helping her son with his school science project as she prepared for her mission.

In orbit, Clark seems to have enjoyed the novelty of moving, eating, sleeping—just living—in zero gravity, and sending messages home: In one e-mail she whe wrote:

> Very busy doing
> science round the clock
> lightning spreading over the Pacific
> the vast plains of Africa
> the dunes on Cape Horn
> rivers breaking through tall mountain passes
> have seen my "friend" Orion several times
> distributed to many
> who I know and love
>
> Hello
> our magnificent planet
> with the cityglow of Australia
> the crescent moon setting
> rivers breaking
> the scars of humanity
> a steep learning curve
> Whenever
> I do get to look out
> lighting up the entire visible horizon
>
> —*The Limb of the Earth Poems By and For Laurel Clark*
>
> *Copied and arranged by*
> *Kezia Vanmeter Sproat, March 29, 2003*

On January 27, eleven days into their flight, the crew paused to commemorate the deaths of Virgil Grissom, Roger Chafee and Edward White on that day in 1967, and the deaths of the *Challenger*'s crew almost exactly seventeen years earlier. Unlike *Challenger,* which never made it into orbit, the *Columbia* crew was enjoying a seemingly perfect mission.

The data stowed, *Columbia* began reentry. At some point in the process, something, perhaps whatever had struck the left wing, destroyed the heat shield. Sixteen minutes before their expected arrival at Cape Canaveral, computers in Houston signaled danger. Sky watchers in California had already noticed fragments of light coming from the orbiter and said that they had waited for, but never heard, the familiar sonic boom of a returning shuttle.

Early Saturday morning, on the first day of February 2003, people in the United States awoke to news that the shuttle *Columbia* had disappeared. A pall moved across the United States and on to Israel and India. The mourners for KC included thousands of women in India as well as Indian-Americans whose messages filled newspapers and Web sites, praising her for showing that an Indian woman could fulfill her dreams.

There was deep sadness, but not the shock that had stunned the world when *Challenger* exploded. Except for the guest astronaut, Colonel Ilan Ramon, there had been little publicity about the mission. Most Americans were unaware that an orbiting shuttle was about to land. Besides, the public associated space disasters with launches, not landings. In this case, the shattered orbiter rained down over a vast stretch of the Southwest and South. During the weeks that followed, NASA and thousands of volunteers collected pieces of wreckage in Louisiana, Texas and Arizona.

The aftermath, too, differed from the postmortem following the *Challenger* disaster. The first tragedy was unique. In the wake of the second, they became a series of two—and who knew how many more—and the shuttle itself became problematic. An advisory board was set up to investigate the accident and once again Sally Ride, retired from the astronaut corps since 1987 and now a physicist on the faculty of the University of California at San Diego and research director at the California Space Institute, was asked to serve. She was the only person to sit on the boards of both shuttle inquiries. She noticed "a little bit of an echo," she said as the board gathered in Houston. In both tragedies there had been serious problems with a particular component of the orbiter that had worn down after repeated flights—the O-rings in 1986, debris from the shuttle's insulating foam tiles in 2003. Analysts attributed both tragedies to the "normalization of risk," the belief that because there had not been an accident from a familiar fault, there would never be one. Were these fatal miscalculations a result of inadequate budgets, or was something basically awry within NASA's culture? Was each accident separate, the resemblances misleading? Or was the shuttle too complicated, its design and technology too old-fashioned? Could they, should they, continue to fly?

At the memorial service in Houston for the *Columbia* crew, a sailor rang a ship's bell seven times in tribute to the lost astronauts. Still "sailing" in orbit 245 miles above Earth, the crew on the International Space Station rang their ship's bell in response.

The Space Station was a major reason why the aftermaths of *Challenger* and *Columbia* were so different. In 1986, the shuttle fleet could sit on the ground for almost three years while the spacecraft were being evaluated. But the Space Station, the only oasis for humans in orbit, would be hard to maintain without help from a shuttle. Since 1986 the Cold War had ended and the Space Station had become an arena for international cooperation. Even incomplete, it is a springboard into deeper space, the place to investigate the effects of prolonged weightlessness and exposure to different kinds of radiation, and to prepare astronauts for a return to the Moon or a trip to Mars.

Yet budgetary pressures have affected Space Station plans. NASA administrator Sean O'Keefe first froze the Station at a crew of three, then cancelled the order for a $16-million "small-size" space suit that had already been made in sizes medium and large. Without it, many women astronauts would be unable to work there because everyone on the Station has to be able to perform an EVA. The cancellation was largely symbolic. In reality only one or two women a year will live on the Station for the rest of the decade. But by arbitrarily dispensing with the one size used only by women, NASA limited the Station's crews to tall women but all men.

The decision prompted the National Space Biomedical Research Institute, a confederation of scientists who advise NASA, to recall what they had earlier warned about: "a space 'glass ceiling' should not exist based on size or gender." O'Keefe restored equity to the female astronauts a year later with the announcement of an altogether new, one-size-fits-all space suit, in which the smallest woman could do a spacewalk.

Several generations of space suits have come and gone since NASA first sent people into orbit, and NASA has learned that it can make adjustments. Women are now integral to the space program and are likely to continue as central actors in it. Unlike Russia and Europe and Canada, each of which boasts a single woman astronaut, NASA has never been guilty of tokenism. By February 2003, NASA had sent thirty-six women into orbit, and eleven more were awaiting their first crew assignment.

There are now so many women astronauts that few people recognize their names. They are simply ordinary astronauts, a phrase that might once have seemed an oxymoron, but now fits these remarkable women who have triumphed over the double obstacles of gender discrimination and the hazards of spaceflight. Like their male colleagues, they are everyday heroes whose exploits still make the news, but not the front page, except when disaster strikes or a spacecraft lands on the Moon or Mars. They make a living, even as they make history, by exploring phenomena and places where no one has gone before.

...2 John Glenn had testified that it was "just a fact of our social order" ...women stayed home while their men took risks for them. That "fact" has c... nged, as has Glenn's attitude toward women astronauts. There are still far fewer women than men in the astronaut corps, but the percentage of women has doubled since 1978. Indeed, the whole social order has changed. Women in the twenty-first century are free to choose risky occupations like fighting fires, captaining fishing vessels or flying military aircraft. What KC Chawla and Laurel Clark wanted, and won, was the right to decide for themselves what risks they were willing to take. Women are now an important part of the astronaut corps. Their fate is tied to that of the space program, not to the obsolete attitudes that dominated the first years of human spaceflight.

Asked about traveling to Mars, would she have gone? KC answered, "In a heartbeat!" She loved being in space, knew the perils and discounted them. Like the rest of her crew, male and female, she lived a life of adventure and exploration and had a chance to soar through the heavens.

NOTES

I am indebted to a great many people who have agreed to be interviewed for this book. Many astronauts spoke with me and, unless otherwise cited, quotations from astronauts and cosmonauts are from these interviews, which are listed at the beginning of each chapter in which they are featured. In addition, a list of all the astronauts interviewed as well as other people involved with the astronaut or cosmonaut programs can be found at the end of these notes. I was extremely fortunate to find newspaper and magazine clippings as well as press releases waiting for me in the archives of NASA's History Office in Washington, D.C., and in the archives at the Smithsonian National Air and Space Museum. I was also fortunate to have the assistance of the History Office and the librarian in Archives at the Johnson Space Center, where I was able to look through the director's files. Some of these files are online from the Johnson Space Center's Oral History Collection and from the Oral History Collection at the National Air and Space Museum.

PROLOGUE

This chapter includes material from interviews with Eileen Collins and Wally Funk.

CHAPTER 1. ASTRONAUTS AND ASTRONETTES

I wish to thank Margaret Weitekamp for untangling the role of Jackie Cochran and the FLATS in her doctoral dissertation, "The Right Stuff, The Wrong Sex." Her formidable persistence and sleuthing have revealed layers of personal connections and chicanery in the relationships between the key figures in this story.

Page 5 **this information:** Barbara Land, *The New Explorers: Women in Antarctica,* p. 21.

Page 7 **"all astronuats should":** *Washington Post,* June 22, 1983.

Page 7 **the cost per pound:** *Washington Star*, October 31, 1958.

Page 7 **females seemed to endure:** *Philadelphia Inquirer*, January 13, 1958.

Page 8 Personal communication from Susan Kleinberg, November 2003.

Page 8 **WISE:** Margaret Weitekamp, p. 159.

Page 9 **Only evidence of the tests:** Ibid., p. 160.

Page 11 **Space and Naval Medicine: Congress in Stockholm:** Joseph D. Atkinson and Jay M. Shafritz, *The Real Stuff*, p. 88.

Page 11 **"We have been informed":** Leslie Haynsworth and David Toomey, *Amelia Earhart's Daughters*, pp. 215–216.

Page 12 **"Monthly physiologic changes":** Johnnie R. Betson, Jr., and Robert R. Secrest, "Prospective Women Astronauts Selection Program."

Page 13 **She addressed them as:** Weitekamp, p. 273.

Page 13 **Gilruth delayed:** letter from Robert R. Gilruth to Jerrie Cobb, April 17, 1962.

Page 14 **scrawled in large letters:** unmailed letter from Lyndon Johnson to James Webb, March 15, 1962, and from Wernher von Braun to Jerrie Cobb, May 19, 1962.

Page 16 **Cobb testified first:** U.S. Congress, House Committee on Science and Astronautics, *Qualifications for Astronauts,* p. 9.

Page 16 **"I think there are many":** Weitekamp, p. 306.

Page 16 **NASA presented its case:** U.S. Congress, House Committee on Science and Astronautics.

Page 17 **had to wait for the laughter to subside:** *Washington Post*, July 19, 1962.

Page 17 **The next to speak was Scott Carpenter:** Chris Kraft, *Flight: My Life in Mission Control*, pp. 162–170.

Page 18 Andy Chaikin, **"That's why we're doing it":** "Tape Reveals JFK Clash with NASA over Moon Program." *SPACE NEWS,* August 27, 2001. p. 4.

CHAPTER 2. TWO VALENTINAS

For this and Chapter 5, I am indebted to James Harford's *Korolev* and Asif A. Siddiqi's *Challenge to Apollo: The Soviet Union and the Space Race, 1945–1974.* I am grateful to Slava Gurevitch, who not only helped translate interviews taped in Moscow, but did his own interview of Valentina Ponomareva and translated parts of her memoir, which he has put on the Internet. In Moscow I was also helped by the staff at Video Cosmos and the medical-psychological team at the Gagarin Center in Star City, and the staff at the IBMP Medical Center in Moscow. They answered my questions with candor. Some of their names are acknowledged elsewhere, but they are very much a part of these notes.

I am also grateful for interviews with Valentina Ponomareva, Irena Solovieva and Zhanna Yerkina.

Page 20 **"It was a distinct trait":** Author interview with Valentina Ponomareva.

Page 20 **Constitution of the Russian Soviet: Federated Social Republic:** 1918 Constitution of the Russian Soviet Federated Social Republic. Adopted by the fifth

All-Russian Congress of Soviets, July 10, 1918. Article Four, THE RIGHT TO VOTE.

Page 22 **"First, inevitably, the idea":** K. E. Tsiolkovskii, "Exploration of the Universe with Reactive Devices," in *Collected Works of K. E. Tsiolkovskii, Vol. 2, Reactive Flying Machines.*

Page 22 **Korolev escaped execution not arrest:** James Harford, *Korolev,* pp. 50–57.

Page 25 **"A man, then men":** Personal conversation with Sergei Khrushchev, by telephone, August 1999.

Page 25 **women will definitely fly:** Nikolai Kamanin diaries as quoted in Valentina Ponomareva and Debra Facktor, "The Flight That Never Happened"

Page 26 **that would include five women:** Slava Gurevitch interview with Ponomareva, May 21, 2002, http://hrst.mit.edu/hrs/apollo/soviet/interview-ponomareva.htm.

Page 26 **Ponomareva thought her chances:** Ibid.

Page 27 **We know from the diaries:** *Kamanin Diaries.*

Page 27 **"The very nature of communism":** Ponomareva, *The Female Face of the Cosmos.*

Page 29 **"You will dive from this ship":** Author interviews with Zhanna Yerkina and Irina Solovieva, both on August 18, 1999.

Page 30 **"Gagarin in a skirt":** Kamanin, *Skrytiy Kosmos, 1960–1963,* cited in Siddiqi, *Challenge to Apollo,* p. 262.

Page 32 **Tereshkova insisted she felt fine:** Mitchell Sharpe, *It is I, Seagull,* and Siddiqi, *Challenge to Apollo,* pp. 361–362.

Page 34 **the Chief Designer swore:** Siddiqi, *Challenge to Apollo,* p. 372.

Page 34 **"how to deceive physicians":** Gurevitch interview with Ponomareva.

Page 34 **In an interview in 1999:** Author interview with Dr. Olga Kozerenko.

Page 36 **Hart, a more acerbic personality:** *New York Herald Tribune,* June 17, 1963.

Page 36 **"The Soviet Union has given...":** The *New York Times,* June 17, 1963.

Page 37 **"Once again the Russians . . .":** *Life,* June 28, 1963, p. 31.

Page 37 **"makes me sick at my stomach":** Sue Solet, *New York Herald Tribune,* June 17, 1963, p. 1.

CHAPTER 3. STILL GROUNDED

Constance Penley's *NASA/Trek: Popular Science and Sex in America*, and Penley's own account of her experience with NASA, were especially helpful with this chapter. Interviews with Mary Connor, Patricia Cowings, Emily Holton and Ann Whitaker helped bring this era into focus.

Page 39 ***Barbarella,* adapted from:** Written by Jean-Claude Forest, 1962.

Page 41 **Janette Piccard, a famous Belgian:** David DeVorkin, *Race to the Stratosphere,* p. 320.

Page 42 **"understanding husbands"** and **"If a woman scientist":** Marion Dietrich, "Role of Women in Space Questioned," *Oakland* (California) *Tribune,* October 29, 1966, pp. 4–5.

Page 42 **Jack Anderson, the columnist who:** Jack Anderson, "Would-be Astronauts: Legion of Angry Women." *Parade,* November, 19, 1967.

Page 43 **women tend to get married:** *San Diego Union,* March 7, 1968.

Page 45 **For the first time Barbie:** Personal communication with Lisa McKendall, corporate communications person at Mattel, March 12, 1999, http://www.ibiblio.org/stayfree/11/barmie.htm.

Page 45 **Charles Berry:** *Miami Herald,* May 21, 1972.

Page 45 **"the possibilities of weightlessness":** Michael Collins, *Carrying the Fire: An Astronaut's Journey,* p. 360.

Page 46 **"only the moon":** Isaac Asimov, *Ladies Home Journal,* February 1971

Page 46 **"Creature comforts. Period.":** *Miami Herald,* May 21, 1972.

Page 47 **"ever found a woman capable":** *Boston Globe,* February 23, 1971.

Page 47 **"We have no plans to send any women":** *New York Times*, March 2, 1972.

Page 49 **Nixon had no choice:** Joan Hoff, "The Presidency, Congress, and the Deceleration of the U.S. Space Program in the 1970s," in Roger D. Launius and Howard E. McCurdy, eds., *Spaceflight and the Myth of Presidential Leadership,* p. 109.

Page 50 **"it was generally accepted [that] women":** Author interview with Bob Parker, March 1999.

Page 51 **When they got to the drawing board:** On August 16, 1972, NASA issued a press release about "BATHROOM COMMODE DESIGN FOR SPACE SHUTTLE PASSENGERS." The release concluded with the statement that Hamilton Standard was building a prototype under a contract to NASA for $238,000. This was referred to as a "Space Powder Room" in *Aviation Week* on August 14, 1972. These documents are in files at the Smithsonian National Air and Space Museum in Washington, D.C. Seven years later the matter appears in a letter to Chris Kraft at the Johnson Space Center from Texas congressman Ron Paul, on October 19, 1979, with a long description of a "Female Urine Management Device" (found in the Directors file at JSC). On April 6, 1972, Charles A. Berry, NASA Director for Life Sciences, sent out a "Request for Female Subjects to Surport Og [*sic*] Testing of Prototype Shuttle Washte Management System."

Page 52 **"Four Wives" had been named:** *Huntsville (Alabama) Times*, September 18 and December 23, 1977.

Page 53 **"the best damn animal":** Author interview with Patricia Cowings, October 12, 2000.

CHAPTER 4. EXPLORERS OR PIONEERS

Seven of the first eight women astronauts spoke with me for this chapter: Rhea Seddon, Kathy Sullivan, Anna Fisher, Sally Ride, Shannon Lucid, Mary Cleave and Bonnie Dunbar. Marvin Resnik spoke to me about his daughter Judy.

Page 56 **"Now we're sending up":** *Geo,* November 1982, p. 21.

Page 58 **There have been three females:** *Congressional Record,* December 4, 1974.

Page 59 **"Later that year":** Harris, *Washington Post,* June 22, 1983, p. A1.

Page 59 **"the subject of including women never came up":** Kraft, *Flight: My Life in Mission Control,* p. 85.

Page 60 **"No one could find":** Author interview with Carolyn Huntoon.

Page 61 **"Where the hell is":** Nichelle Nichols, *Beyond Uhura,* pp. 210–211.

Page 63 **"good cop, bad cop":** *MS,* January 1983, p. 48.

Page 63 **the road less traveled:** *Geo,* September 1982.

Page 64 **"They didn't have to be great scientists":** Author interview with George Abbey, March 29, 2000.

Page 64 **"two is not enough":** Personal communication. Tommy O'Toole, 2001.

Page 64 **NASA had the idea that these brave, white:** Mae Mills Link, *Space Medicine in Project Mercury,* p. 47.

Page 68 **"mostly let them":** Janet Wiscombe, "Private Property," p. 35.

Page 76 **Brian Duff:** May 24, 1989, Oral History, John Mauer, May 124, 1989, The Glennan–Webb–Seamans Project for Research in Space History, National Air and Space Museum.

Page 76 **"the other Sally":** Lydia Dotto, *Canada in Space,* p. 127.

Page 77 **"an all-female crew":** *Washington Times,* March 8, 1983.

CHAPTER 5. BACK IN STAR CITY . . .

Cosmonauts Irina Pronina and Natalya Kuleshova provided a joint interview at the offices of Energia in Korolev. I also spoke to psychiatrist Steve Rostislav Bogdashevsky in Star City and to Victor Baranov and Olga Kozerenko at the Biomedical Space Institute in Moscow. Catherine Lewis, curator at the National Air and Space Museum, directed me to many useful documents.

Page 81 **"shrink heart size":** B. J. Bluth, "Soviet Space Stress," *Science 81,* September.

Page 83 **Tereshkova is a national treasure:** This passage and discussion of Glushko is from an essay written in honor of Valentin Glushko by Svetlana Savitskaia, in 1998. Asif A. Siddiqi sent it to me and Slava Gurevitch provided me with a translation.

Page 84 **Glushko had to correct:** *Spaceflight,* Vol. 41 (November 1999), pp. 475–476.

Page 85 **The only daughter:** Gordon R. Hooper, *The Soviet Cosmonaut Team,* 1986.

Page 86 **an apron ready:** *New York Times,* August 29, 1982.

Page 87 **Leningrad psychologist, V. Garbuzuv:** Seth Mydans, "Place for Women in Space Asserted," *New York Times,* August 11, 1983.

Page 87 **Serebrov was antagonistic:** *Spaceflight, Vol. 41,* p. 476.

Page 87 **a superstition among sailors:** Ibid.

Page 87 **Giving a girl an apron:** New York Times, August 11, 1984.

Page 88 **ever cheerful and lie a lot:** Ibid.

Page 88 **the presence of a woman:** *New York Times,* August 29, 1982.

Page 89 **This trip would be a double strike:** *Spaceflight, Vol. 41*, p. 478.

Page 90 **has vividly shown a possibility of women's effective activity:** Ibid., p. 479.

Page 90 **"We do not go into space":** *New York Times,* August 11, 1984.

Page 90 **the smells rushed in:** Ibid.

CHAPTER 6. TWO AMERICAN WOMEN

Marvin Resnik spoke to me and provided me with articles written about his daughter. I found Robert T. Holder's *I Touch the Future . . .* especially useful in writing about Christa McAuliffe.

Page 94 **A decade earlier:** Diane Vaughan, *The Challenger Launch Decision,* p. 24.

Page 96 **She let her hair grow:** Barbara Galloway, "A Private Astronaut."

Page 97 **"Angel, I am here":** *Washington Post* Staff, *Challengers,* p. 86.

Page 97 **"It's the kind of question":** Ibid., p. 91.

Page 99 **"she had found":** Ibid., p. 83.

Page 99 **"What do you say":** Ibid., p. 152.

Page 99 **"I've been the only woman":** Lon Rains, "Astronaut Lauds Another Woman Pioneer," *Washington Post,* September 25, 1983.

Page 100 **"The only thing":** *Challengers,* p. 152.

Page 102 **"As a woman":** Ibid., p. 158.

Page 102 **"NASA wasn't looking":** Holder, *I Touch the Future . . .* , p. 66.

Page 102**"It was strange, really":** Ibid., p. 68.

Page 103 **"a very diverse group of people":** Ibid., p. 83.

Page 104 **"Only the men get sick":** Ibid., p. 109.

Page 106 **"What are we going":** Ibid., p. 113.

Page 106 **"My old math teacher":** Barbara Galloway, "Akron's Astronaut," *Akron Beacon,* p. 10.

Page 107 **"I'll eat it when I get back":** Holder, *I Touch the Future . . .* , p. 249.

Page 108 **Martin Harwit:** Personal communication.

Page 112 **My daughter Christa McAuliffe:** Grace George Corrigan, *A Journal for Christa,* p. 6.

Page 112 **"No one in big business":** http://www.lifemag.com/life/space/challenger/challenger05.html.

Page 112 **"No," not as things were:** *Houston Chronicle,* September 21, 1987.

Page 112 **Ride Report:** Sally K. Ride, *Leadership and America's Future in Space.*

CHAPTER 7. FROM THE ASHES

Mae Jemison, Millie Hughes-Fulford and Chiaki Mukai provided lengthy interviews. Most of the information on Roberta Bondar is from her memoir, and from material in the NASA's archives.

Page 115 **astronaut program in the American psyche. Each lost astronaut had an asteroid named for her:** Asteroid 3356 is Resnik, discovered February 6, 1984, and Asteroid 3352 is McAuliffe, discovered February 6, 1981.

Pages 117 **Public Opinion:** Jon D. Miller, "Impact of the Challenger Accident on Public Attitudes."

Page 123 **"space helmet":** Roberta Bondar, *Touching the Earth,* pp. 10–12.

Page 123 **"have to force the system":** Dotto, *Canada in Space,* p. 15.

Page 123 **"not suffering fools":** John Colapinto, "High-Flying Roberta Bondar."

Page 124 **"In Canada, her home town":** Kellie Hudson, "Hometown Gets Boost from 'Bondar-Mania,'" *Toronto Star,* January 30, 1992, p. 43.

Page 127 **"How does it feel":** Mark Francis Cohen, "Sky's the Limit."

Page 128 **"over and over in zero gravity" (footnote):** http://jinjapan.org/kids web/news/99–03/poem.html.

Page 130 **"aged twenty and 5'9"**: Mae Jemison, *Find Where the Wind Goes,* p. 135.

Page 131 **"Huntsville, Endeavor":** Ibid., p. 177.

Page 133 **"first African-American":** Ibid., p. 194.

Page 133 **Pratiwi Sudarmono is:** Colin Burgess, "Lost Mission."

CHAPTER 8. NOT QUITE A HOTEL

This chapter includes excerpts from an interview with Elena Kondakova; most of the information about Helen Sharman is from her memoir *Seize the Moment.*

Page 138 **"they had never seen":** Bryan Burrough, *Dragonfly,* pp. 60–61.

Page 138 **"Astronaut wanted":** Helen Sharman and Christopher Priest, *Seize the Moment,* pp. 69–70.

Page 141 **"It is not a woman's":** *Spaceflight,* May 2002, p.204–207.

Page 142 **"is demanding work, even for men":** "Soviet Space Strictly Men's Work," *Space News,* April 23–29, 1990, p. 2.

Page 144 **"Quite simply":** Sharman and Priest, *Seize the Moment,* p. 173.

Page 148 **Ryumin maintained:** *Space Today,* May 14, 1997, and *Orlando Sentinel,* May 3, 1997.

Page 150 **"As we are bringing":** Burrough, *Dragonfly,* p. 31.

Pages 150 **"All of a sudden":** Author interview with James Wetherbee, March 28, 2000.

CHAPTER 9. CHATTERBOX IN ORBIT

In this chapter there are excerpts from author interviews with Bonnie Dunbar, Wendy Lawrence and Shannon Lucid, as well as excerpts from oral interviews about Shuttle–*Mir* from the Johnson Space Center, and references to Bryan Burrough's excellent account of *Mir* in *Dragonfly.*

Page 154 **"female urine":** Burrough, *Dragonfly,* p. 284.

Page 155 **"looked beautiful"** and **"It's not my job":** Ibid., p. 285.

Page 155 **"I saw no women":** Bonnie Dunbar, Oral History, Shuttle–*Mir*, Johnson Space Center, 1998.

Page 156 **"Bonnie just brought over":** Burrough, *Dragonfly,* pp. 285–286.

Page 156 **Yuri Kargapolov, "got up":** Dunbar, Oral History, Johnson Space Center, p. 22.

Page 156 **Complete now, with its solar panels…:** this image is from Burrough, *Dragonfly*.

Page 157 **"women love to clean":** Burrough, *Dragonfly*, p. 284.

Page 157 **"For a scientist who":** Shannon W. Lucid, "Six Months on *Mir*," p. 2.

Page 158 **"Your brother's OK":** Burrough, *Dragonfly*, p. 331.

Page 158 **"There was no contact":** Ibid.

Page 158 **"had had a good career":** Ibid., p. 333.

Page 159 **"launched into an":** Ibid., p. 333.

Page 159 **"They acted very happy":** Shannon Lucid, Oral History, Johnson Space Center, pp. 5–6.

Page 160 "**Yuri and Yuri were absolutely**"**:** Ibid.

Page 160 **"I had spent a":** Lucid, "Six Months on *Mir*," p. 1.

Pages 161 **"During some of the experiments":** Ibid., p. 4.

Page 162 **"We never spoke English"** and "**Now you guys"** and **"Well, I think when I look back":** Lucid, Oral History, Johnson Space Center, pp. 21–22.

Page 166 **"enough eggs":** "Des heures supplémentaires pour Claudie André-Deshays à bord de Mer," Agence France Presse, trans. by the author, August 27, 1996.

Page 166 **"culinary tradition":** Amelie de Tuckhein, "Le festin de Claudie," *Elle*, August 26, 1997.

Page 167 **"there are critics":** James Oberg, *Star-Crossed Orbits*.

Page 169 **"I will never communicate with her again":** Burrough, Dragonfly, p. 337.

Page 169 **"If Clinton knew who":** Burrough, *Dragonfly*, p. 327.

CHAPTER 10. OFFICERS AND GENTLEWOMEN

This chapter includes excerpts from interviews with astronauts Ellen Baker, Catherine Coleman, Eileen Collins, Kay Hire, Susan Kilrain, Wendy Lawrence, Ellen Ochoa, Kathryn Thornton and Shannon Lucid.

Page 186 footnote **William J. Rowe:** W. J. Rowe, "Gender Differences in Astronaut Selection," p. 939; "Optimal Time in the Menstrual Cycle for Spacewalks," pp. 770–771; "Space Flight-Related Endothelial Dysfunction with Potential Congestive Heart Failure," p. 13.

CHAPTER 11. AT HOME IN SPACE

This chapter includes excerpts from interviews with astronauts Susan Helms and Ellen Baker, as well as interviews with NASA chief scientist (2000–2001) Cathie Olsen and former flight surgeon Patricia Santy, and Judith Lapierre, a Canadian volunteer in Moscow.

Page 195 **"equivalent to the volume of":** Todd Halvorsen, "ISS Astronaut Susan Helms."

Page 195 **"sensitivity sessions":** Albert A. Harrison et al. (eds.), *From Antarctica to Outer Space*, p. 85.

Page 195 **"a quick business trip":** Halvorsen, "ISS Astronaut Susan Helms," pp. 1–2.

Page 199 **"stabilizing." A woman:** Mydans, "Place for Women in Space Asserted."

Page 200 **"I am for an all-female crew":** Author interview with Kathie Olsen.

Page 202 **The single exception seems to be pilots:** Pat Dasch, "Long-Duration Spaceflight," p. 33.

Page 202 **The first female candidates:** John R. Ball and Charles H. Evans (eds.), *Safe Passage,* p. 58.

Page 203 **"all the conditions for murder are met":** From an unpublished manuscript, "175 Days in Space: A Russian Cosmonaut's Private Diary," edited and translated by H. Gris, as noted in Jack Stuster, *Bold Endeavors: Lessons from Polar and Space Exploration* (Annapolis: Naval Institute Press, 1996).

Pages 204 **became agitated and interrupted:** Ibid., p. xvi.

Page 205 **called "capricious":** Patricia Santy, personal communication.

Page 205 **"if you suck up, you go up":** Burrough, *Dragonfly*, p. 22.

Page 208 **It flared again in 1972:** *Los Angeles Times*, March 5, 1972.

Page 208 **might need a sexual outlet:** *Time*, October 9, 1972.

Page 208 **quarters for intimate behavior:** Record #006791 LEK 8/10/6, Sex in Space file in the History Archives at NASA headquarters.

Page 209 **"It is remarkable":** R. J. Levin, "The Effects of Space Travel on Sexuality and the Human Reproductive System," pp. 378–382.

Page 210 **"one partner to exercise control":** Nancy McCullough, "Sex in Space."

Page 210 **Zero-g Sex:** Copyright on cassette to Firebird Records.

Page 211 **All float with the same delicacy:** Michael Collins, *Liftoff,* p. 267.

Page 213 **"a large percentage of that research":** "Space Research May Benefit Women's Health," *Station Break,* January 1994, p. 3.

Page 215 **"A paucity of useful clinical data":** Ball and Evans, *Safe Passage,* p. 20.

Page 216 **"retrograde menstrual flow":** Ibid., p. 58.

EPILOGUE

Excerpts from an author interview with Kalpana Chawla helped with this section.

Page 222 **At the memorial service:** Lee Hockstader, *Washington Post*, February 2, 2003 p. A9.

Page 223 **cancelled the order for a $16 million dollar space suit:** "NASA Decision Not Suited for Women," *Science*, Vol. 295, March 1, 2002, p. 633.

Page 224 There are still far fewer women: The numbers vary by class. In 1990 four women entered in a class of twenty-three. There were three women in the next group in 1992. In 1995 and 1996, women comprised 25 percent of these large classes. Since then the numbers have declined, with three women in twenty-six in the class of 1998, and two in seventeen in the last group selected in 2000. These numbers are small and the fluctuations may be meaningless. Overall, women comprise about 20 percent of active astronauts.

Astronauts Interviewed (Women)
Ellen Baker, Kalpana Chawla, Mary Cleave, Catherine Coleman, Eileen Collins, Jan Davis, Bonnie Dunbar, Anna Fisher, Linda Godwin, Susan Helms, Kay Hire, Millie Hughes-Fulford, Mae Jemison, Tamara Jernigan, Janet Kavandi, Susan Kilrain, Wendy Lawrence, Shannon Lucid, Pamela Melroy, Barbara Morgan, Chiaki Mukai, Ellen Ochoa, Sally Ride, Rhea Seddon, Kathryn Sullivan, Kathryn Thornton, Janice Voss, Mary Ellen Weber, Stephanie Wilson.

Astronauts Interviewed (Men)
Kenneth D. "Taco" Cockrell, Hoot Gibson, Steven Hawley, Robert Parker, Charles Precourt, Jim Wetherbee, John Young.

Cosmonauts Interviewed
Claudie Haigneré, Elena Kondakova, Natalie Kuleshova, Nadezhda Kuzhelnaya, Valentina Ponomareva, Irina Pronina, Irina Solovieva, Zhanna Yerkina.

Additional Interviews
George Abbey, Daniel Goldin, Carolyn Huntoon, Judith Lapierre, Marvin Resnik, Duane Ross, Patricia Santy, Joan Vernikos.

BIBLIOGRAPHY

The published material on women in the space program is scattered throughout the literature on NASA, but with the exception of a few articles and books written by the astronauts themselves, the only episode covered seriously is the story of the FLATS. However, several books do hold a lot of useful information. *Who's Who in Space* by Michael Cassutt is an indispensable reference work. Bryan Burrough's *Dragonfly,* James Harford's biography of Sergei Korolev and Asif Siddiqi's *Challenge to Apollo* provide excellent information about the Soviet space program. NASA has many Web sites that contain astronaut biographies as well as transcribed interviews with selected individuals. NASA's history office is open and accommodating and its files are rich in newspaper and magazine clippings. Each NASA center has its own archive; those at the Johnson Space Center are not easily accessible but worth the effort. The archives at the National Air and Space Museum do not replicate those at NASA and are also a valuable source of information. I used all of the following books and articles to some degree.

BOOKS

Allen, Joseph P., with Russell Martin. *Entering Space*. New York: Stewart, Tabori & Chang, 1984.

André-Deshays, Claudie, and Yolaine De La Bigne. *Une Francaise Dans L'Espace*. France: Plon, 1996.

Atkinson, Joseph D., Jr., and Jay M. Shafritz. *The Real Stuff.* New York: Praeger Scientific Press, 1985.

Ball, John R., and Charles H. Evans, eds. *Safe Passage*. Washington, D.C.: National Academy Press, 2001.

Bondar, Roberta. *Touching the Earth*. Toronto: Key Porter Books, 1994.

Burgess, Colin. *Teacher in Space*. Lincoln: University of Nebraska Press, 2000.

Burrough, Bryan. *Dragonfly*. New York: Harper Collins, 1998.

Burrows, William E. *Exploring Space*. New York: Random House, 1990.

Byerly, Radford, Jr. *Space Policy Alternatives*. Boulder, Colo.: Westview Press, 1992.

Cassutt, Michael. *Who's Who in Space, International Space Station Edition*. New York: Macmillan Library Reference USA, 1999.

Clark, Laurel. "The Limb of the Earth: Three Poems by and for Laurel Clark," copied and arranged by Kezia Vanmeter Sproat, in *Write Now! 28th Anniversary Reading by Members of the Women's Poetry Workshop*. Columbus, Ohio: Northwood ART*Space,* 2003.

Clark, Phillip. *The Soviet Manned Space Program*. New York: Orion Books, 1998.

Collins, Michael. *Carrying the Fire: An Astronaut's Journey*. New York: Ballantine, 1974.

______. *Liftoff: The Story of America's Adventures in Space*. New York: Grove Press, 1988.

Connors, Mary M., Albert Harrison and Faren R. Akins. *Living Aloft*. Washington, D.C.: NASA, 1985.

Corrigan, Grace George. *A Journal for Christa*. Lincoln: University of Nebraska Press, 1993.

Cortright, Edgar M., ed. *Apollo Expeditions to the Moon*. Washington, D.C.: NASA, 1975.

Crouch, Tom D. *Aiming for the Stars*. Washington, D.C.: Smithsonian Institution Press, 1999.

Dallek, Robert. "Johnson, Project Apollo, and the Politics of the Space Program," in Roger D. Launius and Howard E. McCurdy, eds., *Spaceflight and the Myth of Presidential Leadership*. Chicago: University of Chicago Press, 1997.

Davis, Flora. *Moving the Mountain*. New York: Simon & Schuster, 1991.

DeVorkin, David H. *Race to the Stratosphere*. New York: Springer, 1989.

Dotto, Lydia. *Canada in Space*. Toronto: Irwin Publishing, 1987.

______. *The Astronauts: Canada's Voyageurs in Space*. Toronto: Stoddart, 1993.

Freeman, Marsha. *Challenges of Human Space Exploration*. New York: Springer-Verlag, 2000.

Gagarin, Yuri. *Road to the Stars: Notes by Soviet Cosmonaut no. 1,* as told to N. Denisov and S. Buzenko, translated by G. Hanna and D. M'yshne. Moscow: Foreign Languages Publishing House, 1962.

Godwin, Robert, ed. *Apollo 8: The NASA Mission Reports*. Burlington, Ontario, Canada: Apogee Press, 1971.

Harford, James. *Korolev*. New York: John Wiley & Sons, 1997.

Harrison, Albert A. *Spacefaring: The Human Dimension*. Berkeley: University of California Press, 2001.

Harrison, Albert A., Yvonne Clearwater and Christopher McKay, eds. *From Antarctica to Outer Space*. New York: Springer-Verlag, 1991.

Hart, Douglas. *The Encyclopedia of Soviet Spacecraft*. New York: Exeter Books, 1987.

Harvey, Brian. *The New Russian Space Programme*. New York: John Wiley & Sons, 1996.

Haynsworth, Leslie, and David Toomey. *Amelia Earhart's Daughters*. New York: William Morrow & Company, 1998.

Heppenheimer, T. A. *Countdown*. New York: John Wiley & Sons, 1997.

Hoff, Joan. "The Presidency, Congress, and the Deceleration of the U.S. Space Program in the 1970s," in Roger D. Launius and Howard E. McCurdy, eds., *Spaceflight and the Myth of Presidential Leadership.* Chicago: University of Chicago Press, 1997.

Holder, Robert T. *"I Touch the Future . . . ."* New York: Random House, 1986.

Hooper, Gordon R. *The Soviet Cosmonaut Team.* Woodbridge, England: GRH Publications, 1986.

Jemison, Mae. *Find Where the Wind Goes.* New York: Scholastic Press, 2001.

Kamanin Diaries, 1960–1963, translated by Bart Hendricky. *Journal of the Interplanetary Society,* 50, No. 1 (January 1997), pp. 33–40..

Kelley, Martha J. M. "Gender Differences and Leadership: A Study." Alabama: Air War College/Air University, Maxwell Air Force Base, 1997.

Kessler-Harris, Alice. *Out to Work.* Oxford: Oxford University Press, 1982.

Kraft, Chris. *Flight: My Life in Mission Control.* New York: Dutton, 2001.

Land, Barbara. *The New Explorers: Women in Antarctica.* New York: Dodd, Mead & Company, 1981.

Launius, Roger D., and Howard E. McCurdy, eds. *Spaceflight and the Myth of Presidential Leadership.* Chicago: University of Chicago Press, 1997.

Linenger, Jerry M. *Off the Planet.* New York: McGraw-Hill, 2000.

Link, Mae Mills. *Space Medicine in Project Mercury.* Washington, D.C.: NASA, 1965.

Logsdon, John M. *Together in Orbit.* Monographs in Aerospace History #11. Washington, D.C.: NASA, 1998.

Logsdon, John M., with Linda J. Lear, Jannelle Warren-Findley, Ray A. Williamson and Dwayne A. Day, eds. *Exploring the Unknown, Volume I: Organizing for Exploration.* Washington, D.C.: NASA, 1995.

Logsdon, John M., with Dwayne A. Day and Roger D. Launius, eds. *Exploring the Unknown, Volume II: External Relationships.* Washington, D.C.: NASA, 1996.

Logsdon, John M., with Roger D. Launius, David H. Onkst and Stephen J. Garber, eds. *Exploring the Unknown, Volume III: Using Space.* Washington, D.C.: NASA, 1998.

Logsdon, John M., with Ray A. Williamson, Roger D. Launius, Russell J. Acker, Stephen J. Garber and Jonathan L. Friedman, eds. *Exploring the Unknown, Volume IV: Accessing Space.* Washington, D.C.: NASA, 1999.

Lord, M. G., Forever Barbie—*The Unauthorized Biorgraphy of a Real Doll,* New York, Morrow, 1994.

Lothian, A. *Valentina: First Woman in Space.* Edinburgh: The Pentland Press, 1993.

Maier, Pauline, Merrit Roe Smith, Alexander Keyssar and Daniel J. Kevles. *Inventing America: A History of the United States.* New York: W. W. Norton & Company, 2003.

McDonough. Thomas R. *Space: The Next Twenty-Five Years.* New York: John Wiley & Sons, 1987.

Murray, Bruce. *Journey into Space.* New York: W. W. Norton & Company, 1989.

Nichols, Nichelle. *Beyond Uhura.* New York: G. P. Putnam's Sons, 1994.

Nicogossian, Arnauld, Carolyn Leach Huntoon and Sam L. Pool, eds. *Space Physiology and Medicine,* 3rd edition. Philadelphia: Lea & Febiger, 1994.

Noonan, Raymond J. "A Philosophical Inquiry into the Role of Sexology in the Space Life Sciences Research and Human Factors Considerations for Extended Spaceflight." Ann Arbor: University of Michigan Dissertation Services, 1998.

Oberg, Alcestis R. *Spacefarers of the '80s and '90s*. New York: Columbia University Press, 1985.

Oberg, James. *Star-Crossed Orbits*. New York: McGraw-Hill, 2002.

Penley, Constance. *Nasa/Trek: Popular Science and Sex in America*. New York: Verso, 1997.

Perone, Karen. *Life Has Become More Joyous, Comrades*. Bloomington: Indiana University Press, 2000.

Phillip, J. Alfred. *They Had a Dream*. Novato, Calif.: Presidio Press, 1994.

Pitts, John A. Th*e Human Factor*. Washington, D.C.: NASA, 1985.

Roland, Alex. *A Spacefaring People*. Washington, D.C.: NASA, 1985.

Rumerman, Judy A. *Chronology of Space Shuttle Flights, 1981–2000*. Washington, D.C.: NASA, 2000.

Santy, Patricia A. *Choosing the Right Stuff*. Westport, Conn.: Praeger Press, 1994.

Sharman, Helen, and Christopher Priest. *Seize the Moment*. London: Victor Gollancz, 1993.

Sharpe, Mitchell. *"It is I, Seagull." A Biography of Valentina Tereshkova*. New York: Thomas Y. Crowell Company, 1975.

Shayler, David J. *Disasters and Accidents in Manned Spaceflight*. New York: Springer-Verlag, 2000.

Siddiqi, Asif A. *Challenge to Apollo: The Soviet Union and the Space Race, 1945–1974*. Washington, D.C.: NASA, 2000.

Stuster, Jack. *Bold Endeavors*. Annapolis, Md.: Naval Institute Press, 1996.

Suvorov, Vladmir, and Alexander Sabelnikov. *The First Manned Spaceflight*. Commack, N.Y.: Nova Science Publishers, 1997.

Swanson, Glen E., ed. "Before This Decade Is Out . . . Personal Reflections on the Apollo Program." NASA History Series. Washington, D.C.: NASA, 1994.

Thomas, Shirley. *Men of Space, Vol. 3*. Philadelphia: Chilton Company, Book Division, 1961.

Trento, Joseph J. *Prescription for Disaster*. New York: Crown, 1987.

Tsiolkovskii, K. E. "Exploration of the Universe with Reactive Devices," in *Collected Works of K. E. Tsiolkovskii*, ed. A. A. Blagonravov, *Vol. 2, Reactive Flying Machines*. Moscow: Isdatel'stvo Akademii Nauk SSSR, 1954.

Vaughan, Diane. *The Challenger Launch Decision*. Chicago: University of Chicago Press, 1996.

Vladimirov, Leonid, trans. by David Floyd. *The Russian Space Bluff*. New York: The Dial Press, 1973.

Washington Post Staff. *Challengers*. New York: Pocket Books, 1986.

Watkins, Elizabeth Siegel. *On the Pill*. Baltimore: Johns Hopkins University Press, 1998.

Weitekamp, Margaret A. "The Right Stuff, The Wrong Sex: The Science, Culture, and Politics of the Lovelace Woman in Space Program." Doctoral dissertation, Cornell University, Ithaca, N.Y., May 2001.

White, Frank. *The Overview Effect,* 2nd edition. Reston, Va.: AIAA, 1999.

Wolfe, Tom. *The Right Stuff.* New York: Farrar, Straus, Giroux, 1979.
Woodmansee, Laura S. *Women Astronauts.* Burlington, Ontario, Canada: Apogee Books, 2002.
Zimmerman, Jean. *Tailspin.* New York: Doubleday, 1995.

MAGAZINES/JOURNALS/NEWSPAPERS

Ackmann, Martha. "Wally Funk Is Still Determined to Get Her Shot in Space." *Houston Chronicle,* February 11, 2000.
Anderson, Jack. "Legion of Angry Women." *Parade,* November 19, 1967.
Associated Press. "Women Demand Astronaut Roles, Court Post, Storm Names Changed." *New Haven Register,* December 8, 1969.
"Attitudes Towards the U.S. Civilian Space Program." Market Opinion Research, 1988.
Betson, Johnnie R., Jr., and Robert R. Secrest. "Prospective Women Astronauts Selection Program." *American Journal of Obstetrics and Gynecology,* 88 (February 1964).
Bluth, B. J. "Soviet Space Stress." *Science,* 81 (September 1981).
Borenstein, Seth. "Husband's Remarks Don't Fly with Cosmonaut Wife." *Orlando Sentinel,* May 3, 1993.
Brett, Abigail. "A Huge Orbiting Barn." *Washington Post,* May 14, 1973. Burgess, Colin. "Lost Mission." *Spaceflight,* 41 (June 1999).
Burgess, Colin, and Francis French. "Only Males Need Apply." *Spaceflight,* 41 (January 1999), pp. 28–31..
Burnham, Darren L. "From Race Tracks to Ground Tracks." *Spaceflight,* 39 (January 1997).
Burns, John F. "An Apron Awaits Soviet Cosmonauts." *New York Times,* August 29, 1982.
Carreau, Mark. "Sally Ride Is Leaving NASA After Making Major Contributions." *Houston Chronicle,* September 21, 1987.
Chaikin, Andy. "Tape Reveals JFK Clash with NASA over Moon Program." *Space News,* August 27, 2001 p. 4.
Cohen, Mark Francis. "Sky's the Limit." *Continental,* May 1999.
Colapinto, John. "High-Flying Roberta Bondar." *Chatelaine,* January 1993.
Congressional Record, 93rd Congress, 1st Session, December 4, 1973, pp. 7760–7762.
Covault, Craig. "Policy and Technology Shape Manned Space Ops." *Aviation Week & Space Technology,* January 8, 2001, pp. 44–48.
Cucinotta, Francis A. "Once We Know All the Radiobiology We Need to Know, How Can We Use It to Predict Space Radiation Risks and Achieve Fame and Fortune?" *Physica Medica,* 17, Supplement 1 (2001).
Cucinotta, F. A., et al. "Space Radiation and Cataracts in Astronauts." *Radiation Research,* 156 (2001).
Dasch, Pat. "Long-Duration Spaceflight." *Ad Astra,* 19, No. 2 (March/April 1997).

Dietrich, Marion. "Role of Women in Space Questioned." *Oakland Tribune*, October 29, 1966.

French, Francis. "First British Ride Into Space." *Spaceflight*, 44, No. 4 (May 2002).

French, Francis, and Wally Funk. "Another Giant Leap." *Spaceflight,* 41, No. 12 (December 1999).

Funk, Wally. "Another Giant Leap." *International Women Pilot,* 25, No. 5 (September/October 1999).

Galloway, Barbara. "A Private Astronaut." *Akron Beacon Magazine*, June 17, 1984.

Gagarin, Yuri. "Woman of Space." *Spaceworld*, September-October 1963, p. 7.

Goldin, Dan. "Female Shuttle Crew Worth Considering." *Space News,* 13, September 6, 1999.

Groopman, Jerome. "Medicine on Mars." *The New Yorker*, February 14, 2000.

Halvorson, Todd. "Mountain Climbing Kondakova Scaled Heights." *Space Today*, May 14, 1997.

______. "ISS Astronaut Susan Helms." *Space Today*, June 17, 2001.

Hamburg, Jay. "Etched Forever." *Orlando Sentinel*, January 28, 1987.

Hendickx, Bart. "The Kamanin Diaries 1960–1963." *Journal of the British Interplanetary Society*, 50, No. 1 (January 1997), pp. 33–40.

Hiassen, Carl. "Fired, Hired, Still Speaking." *Today*, August 1974.

Hockstader, Lee. *Washington Post*, February 2, 2003, p. A9.

Kennedy, Diana. "An Unscripted Life Starring Herself." *New York Times*, May 6, 2001.

Kirkmas, Don. "Women in Space by 1980, NASA Says." *Birmingham (Alabama) Post-Herald*, June 11, 1974.

Koven, Stan. "High Heels for the Big Whirl." *New York Post,* June 17, 1963.

Kraemer, Sylvia K. "Opinion Polls and the US Civil Space Program." *The Capital* (newsletter of the National Capital Section of the American Institute of Aeronautics and Astronautics), 20, No. 8 (September 1992).

Levin, R. J. "The Effects of Space Travel on Sexuality and the Human Reproductive System." *Journal of the Interplanetary Society*, 47 (1989).

Luce, Clare Boothe. "Some People Simply Never Get the Message." *Life*, June 28, 1963.

Lucid, Shannon W. "Six Months on Mir." *Scientific American*, May 1998.

McCullough, Joan. "13 Who Were Left Behind." *MS*, September 1973.

McCullough, Nancy. "Sex in Space." *True News*, November 1992.

Miller, John D. "Impact of the Challenger Accident on Public Attitudes." Public Opinion Research, January 25, 1987.

Miquel, Jaime, and Kenneth Souza. "Gravity Effects on Reproduction, Development and Aging." *Advances in Space Biology and Medicine*, 1 (1991).

Mohanty, Susmita. "Shuttle–Mir Habitability Database." NASA Johnson Space Center, Houston, 1997.

Mydans, Seth. "Place for Women in Space Asserted." *New York Times*, August 11. 1984.

"NASA Decision Not Suited for Women." *Science*, 295 (March 1, 2002), p. 633.

Naunton, Eva. "A Woman in Outer Space? Good Idea—But Still a Joke." *Miami Herald*, May 21, 1972.

"Pingvin (penguin)—3 Muscle and Bone System Loading Suit," technical data sheet from JSC Zvezda, Smithsonian Museum Space History Division, Washington, D.C.

Ponomareva, Valentina, and Debra D. Facktor. "The Flight That Never Happened: The Story of the First Women Cosmonaut Team." Presented at the 47th International Astronautical Congress, October 7–11, 1996.

Rains, Lon. "Astronaut Lauds Another Woman Pioneer." *Washington Post*, September 25, 1983.

Ride, Sally K. "Leadership and America's Future in Space." A Report to the NASA Administrator, August 1987.

Roe, Dorothy. "Girls, Space Travel Might Be Bit Rough." *Orlando Sentinel*, March 18, 1965.

Rowe, W. J. "Gender Differences in Astronaut Selection" (letter). Av*iation, Space, and Environmental Medicine,* 70 (1999).

______. "Optimal Time in the Menstrual Cycle for Spacewalks" (letter). *Aviation, Space, and Environmental Medicine,* 72 (2001).

______. "Space Flight-Related Endothelial Dysfunction with Potential Congestive Heart Failure" (abstract). *Journal of Heart Failure,* 7 (2002).

Sanborn, Sara. "Sally Ride, Astronaut: The World Is Watching." *Ms.*, January 1983.

Santy, Patricia, and David Jones. "An Overview of International Issues in Astronaut Psychological Selection." *Aviation, Space, Science, and Medicine*, October 1994.

Savitskaia, Svetlana. "Odnazhdy I Navegda? Dokumenty I liudi o sozdatatele raketnykh dvidatelei I . . . " ("Once and Forever? Documents and People About the Creator of Rocket Engines and Space Systems, Valentin Petrovich Glushko"), p. 316, Moscow, 1998.

Schuling, Duelof L. "STS-84: Mission to *Mir.*" *Spaceflight*, 39 (September 1997).

Seddon, Rhea. "The Lady Astronaut's Diet." *Weightwatchers*, May 1979.

Shurkin, Joel N. "Women's Lib Reaches New Heights." *Philadelphia Inquirer*, September 24, 1973.

Simpson, Peggy. "Russian Hails Women's Rights." *Philadelphia Inquirer*, June 20, 1975.

Solet, Sue. "Why Woman in Space: U.S. and Soviet Views." *New York Herald Tribune*, June 17, 1963.

Spencer, Scott. "The Epic Flight of Judith Resnik." *Esquire*, December 1986.

Tanner, Henry. "Bykovsky Sought to Continue Trip." *Boston Globe*, February 23, 1971.

Tereshkova-Nikolayeva, Valentina. "From Outer Space by Parachute!" March 30, 1967. File copy, National Air and Space Museum, Smithsonian Institution, Washington, D.C.

______. "Women in the Age of Science and Technology." *Impact of Space on Society,* 20, No. 1 (January-March 1970).

The New York Times, "Space Doctor Sees Woman Astronauts." March 19, 1970.

The New York Times, "Jane Fonda Chides the White House." October 18, 1983, p. 3.

Tuckhein, Amelie de. "Le festin de Claudie." *Elle*, August 26, 1996.
UPI. "6, Including Woman, Are Named Finalists for Seat on the Spacelab." *New York Times*, December 23, 1977.
U.S. Congress, House Committee on Science and Astronautics, *Qualifications for Astronauts: Report of the Special Subcommittee on the Selection of Astronauts*, 87th Congress, 2nd session, July 17–18, 1962.
Vis, Bert. "Soviet Women Cosmonaut Flight Assignments 1963–1989." *Spaceflight*, 41, No. 11 (November 1999), pp. 474–480.
Von Braun, Wernher. Speech given at Mississippi State College, Starkville, November 19, 1962.
"Vostok Flights Widen Soviet Experience." *Aviation Week & Space Technology*, July 22, 1963.
Weitekamp, Margaret A. "The Science, Politics, and Culture of the 'Astronauttes': Examining Gender in Aerospace History." Presented at the American Historical Association Annual Meeting, January 8, 1999, Washington, D.C.
Wilford, John Noble. "Controversy over Jane Fonda Costs NASA Official His Post." *New York Times*, October 15, 1983, p. 6.
Winter, Ruth. "Woman Astronaut? Not Very Soon." *San Diego Union*, March 7, 1968.
Wiscombe, Janet. "Private Property." *Los Angeles Times Magazine*, August 29, 1999.
Witty, Susan. "The First U.S. Women in Space." *Geo*, November 1982, p. 21.

INTERNET

Dunn, Marcia. "Husband thinks a woman's place is on Earth." http://ardemoreite.com/stories/051197/news/news12.html.
Gurevitch, Slava. Interview with Valentina Ponomareva, Moscow, May 21, 2002. Translations from her lecture at the LXV Plenum of Russian National Committee of the History of the Philosophy of Science and Technology, and fragments from the following books: Ponomareva, *The Female Face of the Cosmos* (Moscow: Gelios, 2002); Nikolai Kamanin, *The Hidden Cosmos*, Book 1 (Moscow: Infortekst, 1995).http://hrst.mit.edu/hrs/apollo/soviet/interview/interview-ponomareva.htm.
http://jinjapan.org/kidsweb/news/99–03/poem.html.
http://www.ibiblio.org/stayfree/11/barbie.htm.
http://www.lifemag.com/Life/space/challenger/challenger05.html.

LETTERS

Gilruth, Robert R., to Jerrie Cobb, April 17, 1962. Special Collections, Robert Sherrod Apollo Collection, NASA headquarters History Office, Washington, D.C.

Johnson, Lyndon B., to James Webb, unmailed letter, March 15, 1962. Vice-Presidential Papers, Manuscript Division, Library of Congress, Washington, D.C.

von Braun, Wernher, to Jerrie Cobb, May 19, 1962. Wernher von Braun Papers, Manuscript Division, Library of Congress, Washington, D.C.

MISCELLANEOUS

"Red Space: The Secret Russian Space Program," Show Two of *Untold Stories*, produced by Jones Entertainment Group, Ltd., Robert Fiveson.

The Red Stuff, by Leo de Boer, produced by Pietre van Huystree, The Netherlands, 1999.

Sex in Space file in the History Archives at NASA headquarters, Washington, D.C. Record #006791 LEK 8/10/6.

Shuttle–Mir: The United States and Russia Share History's Highest Stage, a CD by Clay Morgan. Full text of interviews with Russian and American participants in the exchange. Stephen J. Garber and Roger D. Launius, NASA History Office, December 2001.

Women in Space: A Ride to Remember, 1988.

ACKNOWLEDGMENTS

ALTHOUGH THERE IS A LOT of written material about the space programs in the United States and Russia, writing a history of what has just happened, or is still happening, depends inevitably on speaking to people, many of whom helpfully suggest speaking to someone else. To these people I can only say "Thank you." I am also indebted to several institutions for guiding me and providing me with resources to pursue this project.

Starting at the beginning, when the idea took shape in my mind, I found enthusiastic support from John M. Klineberg, former director of the Goddard Space Center in Greenbelt, Maryland. Still feeling my way around, I talked to Charlene Anderson and Lou Friedman at the Planetary Society, who directed me to astronauts on their board, including Kathryn Sullivan. Andy Turnage at the Association of Space Explorers–USA also helped me contact astronauts. At the California Institute of Technology, Annila Sargeant directed me to the astronaut Robert Parker at the Jet Propulsion Laboratory. He provided me with a list of retired women astronauts: Mary Cleave, Kathryn Thornton, Kathryn Sullivan, Rhea Seddon and Mae Jemison, all of whom were frank and invariably fascinating. And then he sent me to meet Carolyn Huntoon, former director of the Johnson Space Center, who discussed with me the range of a possible book, and who I spoke to again when I was midway through the research.

At this point I had intended to write about American women, but the generosity of the Alfred P. Sloan Foundation, to which I am very grateful, enabled me to expand my canvas and include the Soviet program. The Russian adventure could not have been as smooth as it turned out to be without the help of James Harford, who put me in touch with a network of Russians and Americans who were experts in Russian space studies. In Moscow, I thank especially the staff of the Russian journal *I Cosmos*. Through its editor,

Igor Lissov, and with the help of Igor Marinin, I enjoyed the help of Valeriya Davydova, who arranged for me and the historian Valentina Tarasova, who filled in as interpreter at the last minute, to be driven to Star City, Korolev and in Moscow to the Institute of Biomedical Problems. At all these sites, interviews had been arranged in advance with some of the women who had become cosmonauts but did not fly, and with Olga Kozerenko, a physician who had worked with them. They also introduced me to Masha Avdeev, a journalist, whose husband, Sergei Avdeev, was then orbiting the Earth in *Mir.* Through Video Cosmos, they arranged for me to see a program on the astronauts and put me in touch with Robert Fiveson, who sent me an English version. While in Russia I was able to interview Claudie André-Deshays Haigneré, the French cosmonaut, through the assistance of Elizabeth Moussine Pouchkine of CNES, the French space agency. When I returned to the United States I spoke to Sergei Khrushchev, now a professor at Brown University, who had been his father's liaison to the cosmonauts. In St. Petersburg I had come to know Valeria Ivaniushina and Danya Alexandrov, both of whom have served as good sounding boards on Russian matters.

In California I was fortunate in the friendship and assistance of historians Glen Bugos, Peter Neushal and Peter Westwick. I was pleased to meet Mary Connors and Emily Holton at NASA's Ames Center, and Constance Penley at the University of California at Santa Barbara. I am also grateful to Alice Huang at Caltech for her help on women's issues.

At NASA headquarters, when I began my research Arnauld Nicogossian, then head of Life Science Research, explained its history and introduced me to Joan Vernikos, who was very helpful. At headquarters I also received help from Diana P. Hoyt and Victoria Friedensen. There I arranged to interview Daniel Goldin, then administrator of NASA, Kathie Olsen, the chief scientist, and Bert Ulrich, who arranged for me to attend the spectacularly beautiful launch of STS-98 at dusk on February 7, 2001.

In the summer of 2000 I took up an appointment to the Charles A. Lindbergh Chair in Space History at the Smithsonian Institution's National Air and Space Museum. There I found great support and practical advice from curators Cathy Lewis, Valerie Neal, David Dvorkin, Gregg Herkin, Mike Neufeld and Allan Needal. At the museum the librarians were especially helpful, as were the staff in the Archives Division. Not far from the Mall at the History Office at NASA headquarters, Roger Launius, then chief historian, and Nadine Andreassen, Colin Fries, Stephen Garber, John Hargenrader and Jane Odom, the archivist, were as diligent as if they were writing the book themselves. While I was in Washington my agent, Ronald Goldfarb, kindly arranged for me to meet with Roald Sagdeev, former director of the Soviet Space Research Institute and now professor of physics at the University of Maryland.

On visits to the Johnson Space Center in Houston, Texas, historian Glen Swanson led me through their library and directors' files. Also helpful at JSC were Eileen Hawley and Lucy Lytwnsky, who arranged interviews with "flown women" astronauts and some of the "flown" men who were, or had been, heads of the astronaut office. In Houston I was also able to interview Duane Ross and Deborah Harms. At this point Lee Silver, at Caltech, who had taught the Apollo astronauts what to look for on the Moon, arranged for me to interview George Abbey, then director of JSC, and astronauts Jim Wetherbee and John Young.

Research is, of course, a treasure hunt, with one clue leading to another. Among the treasures of that hunt was Patricia Santy, who I met in Houston and have talked to at her new home in Ann Arbor, Michigan, about medical research and the testing of astronauts. In Houston I also met and learned from James Oberg, who writes about the Soviet/Russian program.

Through an internet chat room I was contacted by the Australian writer Colin Burgess, who sent me several unpublished articles on women cosmonauts, and through him I met Francis French, now in San Diego, a veritable encyclopedia of astronaut information.

The telephone can do what the Internet cannot. By telephone I spoke to Enrico Alleva in Rome about Italian biomedical space research. In the United States, I thank astronauts Hoot Gibson and Joe Allen and David Williams, as well as Janet Kavandi, Sally Ride, Tamara Jernigan, Chiaki Mukai, Catherine Coleman and Mae Jeinison. To this group I will add the scientists Laurence Young, Jay Shapiro and John Dicello. I want to thank Ron Noonan for his insight about sex and NASA; his dissertation is an important contribution to the literature.

I had the good fortune to have research assistance from Catherine Spanbock in Los Angeles and from Tammy Ingram at Yale, where I began teaching in 2000. I was lucky to find two fine transcribers, Kris Nelson and Abby Delman. A virtual writer's group, including Marilyn Nissenson, Mary Ellen Strote and Beth Kevles, read through some of the chapters. At Yale I am grateful for the support of the late Frederic L. Holmes, chair of the section for the history of medicine at the Medical School, and Jon Butler, chair of the history department. I am grateful to Naomi Rogers for directing me to important sources of women's history and to administrative assistance in both places.

The 2002–2003 year at Yale brought me Kathryn Czarnowski as a fact check/research assistant. She worked efficiently and intelligently and always with good cheer. Gathering the countless treasures of research is a group effort. Transforming them into a narrative is a lonely job, but it can be helped, and was helped, by a gifted editor. Through the five years from inspiration to bound book, I have had the good fortune to have worked with Amanda Cook, who offered strong editorial guidance was well as confidence. My husband, Daniel Kevles, gave the manuscript critical readings and offered sustained enthusiasm from the moment I embarked upon this book through its completion. Special thanks to Michael Cassutt for saving me from numerous errors. I am grateful to him and to everyone who enabled me to enter into the world of space travel and brought me home safely.

TIME LINE

DATE	POLITICAL EVENTS	AMERICAN SPACE PROGRAM
1917	Beginning of Russian Revolution	
1918	Provisional government in Russia	
1920		
1945	End of WWII	Werhner von Braun and his team brought to the U.S. from a defeated Germany
1950	Korean War begins	von Braun moves to Redstone Arsenal near Huntsville, Alabama to build the Army's Jupiter missile
1952	Dwight Eisenhower elected president	
1953	Stalin dies; Korean War ends	

USSR/RUSSIAN SPACE PROGRAM	WOMEN'S MILESTONES	CULTURAL EVENTS
	Russian women granted suffrage	
	19th Amendment to U.S. Constitution grants women suffrage	
Sergei Korolev finds what was left behind at Nazi rocket base at Peenemunde.		
	Maria Klenova sent with Soviet scientific team to Antarctica	

DATE	POLITICAL EVENTS	AMERICAN SPACE PROGRAM
1957		
1957–1958 (18 month)	International-Geophysical Year	Explorer satellites discover Van Allen asteroid belts around the Earth; Creation of NASA—National Aeronautic and Space Administration
1959		NASA tests candidates, military test-pilots only, to become astronauts; Mercury Seven selected
1960	John F. Kennedy elected president	
1961	April: Bay of Pigs August: Berlin Wall construction begins; September: Soviets resume nuclear tests May 25: President Kennedy delivers his "in a decade" speech. The space race begins.	May 5: Alan Shepard successfully rockets into suborbit for 15 minutes.

USSR/RUSSIAN SPACE PROGRAM	WOMEN'S MILESTONES	CULTURAL EVENTS
Oct.: *Sputnik* I launched Nov.: Sputnik II with Laika, a dog.	Eisenhower creates Presidential Commission on Women	
	First systematic studies conducted by the US in Antarctica—limited to male scientists	
	Antarctica made international continent for scientific studies.	
cosmonaut selected	Lovelace tests Jerri Cobb in Albuquerque	Birth control pill available commercially
April: Yuri Gagarin becomes first man in space August: German Titov becomes second man in space	Kennedy establishes the Presidential Commission on the Status of Women Lovelace selects 12 other women candidates: the FLATS, retrospectively called the Mercury 13	

DATE	POLITICAL EVENTS	AMERICAN SPACE PROGRAM
1962	October: Cuban Missile Crisis	February: John Glenn becomes first American in orbit April: Scott Carpenter becomes second
1963	November: John F. Kennedy assassinated	
1964	Lyndon Baines Johnson elected President; Civil Rights Act passed in the U.S.; Vietnam War begins	
1965	Vietnam War escalates; Voting Rights Act passed in the U.S.	
1966		
1967		January: *Apollo 1* fire; deaths of White, Grissom, and Chaffee

USSR/RUSSIAN SPACE PROGRAM	WOMEN'S MILESTONES	CULTURAL EVENTS
	July: Subcommittee on the Selection of Astronauts hears American women's case for becoming astronauts	
June: Valentina Tereshkova becomes the first woman and sixth cosmonaut to orbit the Earth		Betty Friedan publishes *The Feminine Mystique*
	"Sex" added to Title VII of the Civil Rights Act	
		Astronaut Barbie introduced
	National Organization of Women established	*Star Trek*'s first season
	First women visit McMurdo Station in Antarctica	

DATE	POLITICAL EVENTS	AMERICAN SPACE PROGRAM
1968	Richard Nixon elected president	
1969		July: Apollo 11 puts Neil Armstrong and Buzz Aldrin on the Moon
1970	July: Apollo 13 rescued	
1971		
1973		First and second *Skylab* missions
1974	Nixon resigns; Gerald Ford becomes president; the U.S. withdraws from Vietnam	*Skylab*'s third and final mission
1976	Jimmy Carter elected president	

USSR/RUSSIAN SPACE PROGRAM	WOMEN'S MILESTONES	CULTURAL EVENTS
		Barbarella and *2001: A Space Odyssey* released; Madeleine L'Engle's *A Wrinkle in Time* published
	Four U.S. women camp near, but not at, Naval Station at McMurdo in Antarctica	
		Our Bodies, Ourselves published by the Boston Women's Health Book Collective
	Roe v. Wade legalized abortion in the U.S.; Navy accepts women pilots	
Apollo–Soyuz link-up: first use of space for diplomacy	Two women scientists winter over at McMurdo Station; Air Force accepts women test pilots	

DATE	POLITICAL EVENTS	AMERICAN SPACE PROGRAM
1978		
1980	Ronald Reagan elected president	
1981	*Columbia*, the first shuttle, is launched	
1982		
1983		Sally Ride becomes first American woman to orbit the Earth
1984	Ronald Reagan reelected	
1986	Glasnost proclaimed by Soviet leader Mikhail Gorbachev	January: *Challenger* explodes after launch July: Rogers Commission begins investigation of the disaster
1988	George H.W. Bush elected president	

USSR/RUSSIAN SPACE PROGRAM	WOMEN'S MILESTONES	CULTURAL EVENTS
	NASA announces first astronaut class with Mission Specialists including six women astronauts; Navy invites women to become equal partners in Antarctic research	
Svetlana Savitskaia flies on Soviet space station *Salyut 7*		
	Sally Ride flies on STS-7	
July: Savitskaia becomes first woman to perform a spacewalk	October: Kathryn Sullivan first American woman to perform a space walk	
Russians launch *Mir* space station	First American women astronauts perish in space: Judith Resnik and Christa McAuliffe	

DATE	POLITICAL EVENTS	AMERICAN SPACE PROGRAM
1990		
1991	Soviet Union collapses	
1992	Bill Clinton elected president	
1994		
1995		Shuttle Discovery rendezvous with *Mir*; Norman Thagard - becomes first U.S. astronaut to live and work with Russians as Shuttle-*Mir* part of cooperation
1996		
1998		John Glenn second flight

USSR/RUSSIAN SPACE PROGRAM	WOMEN'S MILESTONES	CULTURAL EVENTS
Helen Sharman becomes first and only British woman in space as a guest on *Mir*	Eileen Collins: becomes first female pilot astronaut to join NASA	
		Women's Health Initiative begun by U.S. National Institutes of Health
Elena Kondakova flies as first Russian woman on *Mir*		
Claudie André-Deshays (Heigneré) visits *Mir* as first French/European cosmonaut	Shannon Lucid stays on *Mir* for record 188 days	

DATE	POLITICAL EVENTS	AMERICAN SPACE PROGRAM
1999		
2000	George W. Bush becomes president	International Space Station gets first crew
2001	September 11: Terrorists attack World Trade Center in New York City and Pentagon in Virginia	
2003		February: Shuttle *Columbia* disintegrates upon reentry -

USSR/RUSSIAN SPACE PROGRAM	WOMEN'S MILESTONES	CULTURAL EVENTS
	Eileen Collins flies as first female shuttle commander	
Mir retired, burns on reentry into Earth's atmosphere; Budget cuts freeze ISS with capacity to rescue only three astronauts		
	Susan Helms becomes first woman resident of the International Space Station	
	Kalpana Chawla and Laurel Clark perish with the rest of Columbia's seven-person crew	

INDEX